DeWALT®

W9-CKV-513

BLUEPRINT
READING

PROFESSIONAL REFERENCE

Paul Rosenberg

Created exclusively
for DeWALT by:

AUG   2007

PAL
publications®

www.palpublications.com
1-800-246-2175

Titles Available From DeWALT

DeWALT Trade Reference Series

Blueprint Reading Professional Reference
Construction Professional Reference
Datacom Professional Reference
Electric Motor Professional Reference
Electrical Estimating Professional Reference
Electrical Professional Reference
HVAC/R Professional Reference—Master Edition
Spanish/English Construction Dictionary—Illustrated
Lighting & Maintenance Professional Reference
Plumbing Professional Reference
Wiring Diagrams Professional Reference

DeWALT Exam and Certification Series

Building Contractor's Licensing Exam Guide
Electrical Licensing Exam Guide
HVAC Technician Certification Exam Guide
Plumbing Licensing Exam Guide

For a complete list of The DeWALT Professional Trade Reference Series visit **www.palpublications.com**.

This Book Belongs To:

Name: ______________________________

Company: ______________________________

Title: ______________________________

Department: ______________________________

Company Address: ______________________________

____

____

Company Phone: ______________________________

Home Phone: ______________________________

Pal Publications, Inc.
374 Circle of Progress Drive
Pottstown, PA 19464-3810

Copyright © 2006 by Pal Publications, Inc.
First edition published 2006

ISBN 0-9770003-5-4

10   09   08   07   06   5   4   3   2   1

Printed in the United States of America

A Note To Our Customers

We have manufactured this book to the highest quality standards possible. The cover is made of a flexible, durable and water-resistant material able to withstand the toughest on-the-job conditions. We also utilize the Otabind process which allows this book to lay flatter than traditional paperback books that tend to snap shut while in use.

Preface

Blueprint Reading is a necessary skill for every participant in the construction industry. Whether you are a plumber, carpenter, roofer, electrician, flooring installer, or any other trade professional involved in the industry, it is crucial to have the skills necessary to read blueprints.

This book begins with the basic information and skills you need to read and understand blueprints. This material is the same whether you are reading foundation plans, HVAC plans, or any other type of plan. Everyone, regardless of specialty, should give special attention to chapter one, *Blueprint Basics*.

Chapter two and chapter three, *Specifications and Architectural Drawings*, respectively, are also of importance to everyone. *Specifications* compliment almost every blueprint, while *Architectural Drawings* are necessary for almost all other pages in a set of blueprints.

Chapter four through chapter eight cover special types of blueprints used for the primary specialty trades — structural, plumbing, mechanical and electrical.

Chapter nine and ten conclude this book with a detailed glossary and list of abbreviations used in the construction trades.

Naturally, there may be some aspects of blueprint reading that I have not covered in sufficient depth for some readers. I will update this book on a continual basis and will attempt to include material suggested by readers as well as keep pace with developments in the construction industry.

Best wishes,
Paul Rosenberg

CONTENTS

CHAPTER 2 – *Specifications* 2-1

CHAPTER 5 – *Structural Drawings* . . . 5-1

CHAPTER 6 – *Plumbing Drawings* 6-1

CHAPTER 8 – *Electrical Drawings* 8-1

CHAPTER 1
Blueprint Basics

BLUEPRINTS

The first thing to understand is that what we call *blueprints* are usually not blue. Years ago they were blue drawings with white lines. This was a result of the process used to produce them. Now, blueprints are usually white pages with black or blue lines. Nonetheless, the name *blueprint* has remained and probably will remain in use for a long time to come. You may also hear blueprints referred to as *drawings*, *prints*, or *plans*.

A blueprint is a representation of what is to be constructed. It is a drawing of what is to be built. Blueprints, however, are very precise drawings. They are exact representations of what is to be built. Obviously, they are much, much smaller than the proposed structure, but they are exact and detailed.

Every line on a construction drawing is carefully placed. The relation of a line to another line shows distances.

Usually, blueprints display a view of the project as seen from above—in other words, as it might be viewed by a bird. However, many blueprints show a view that cannot be seen in real life. For example, many floor plans show the walls and floor, providing a view that could only be possible in real life if you removed the roof of the completed building. Other blueprints may show only the foundation and floor slab, or a single floor of a multifloor building.

BLUEPRINTS *(cont.)*

Properly, the view from above is termed a *plan*; drawings shown from other perspectives have different names. Engineers may use such terms correctly, but this is seldom true on a construction site. The term *plan* is generally used for any type of drawing.

Blueprints show a great deal of information. You must read them carefully and slowly. If you skim or go too quickly, you will almost certainly miss a number of important (and expensive) items.

BLUEPRINT QUALITY

Be aware that the quality of blueprints used for construction is sometimes poor. The printing may be difficult to read. Or important information may be missing from the drawings. Entire pages may be missing, or you may receive only a set of plans or only a set of specifications.

If your prints are incomplete or of poor quality, be careful. Make sure that the people you work for, whether a contractor, building owner, architect, etc., are clear that you lack the proper documents. If you have to sign any sort of contract, make a note on it, specifying exactly what you are basing the agreement upon. For example:

> "This price is based on pages A1, A2, A3, A4, M1, M2, M3, M4, and E1 provided to me by ABC Contracting. The plan sheets are dated 6/1/2005 and were prepared by James & Co. Architects."

BLUEPRINT QUALITY *(cont.)*

If the prints are not clear enough to read and use, wait until you have better ones or make your own. Otherwise you may encounter serious problems in the course of the project.

Use your judgment well.

HOW TO READ A SET OF BLUEPRINTS

Keep in mind that blueprints come in sets. The set of prints for a single house may contain only a few sheets. A large project, however, may contain a hundred sheets.

The general process for reading blueprints is as follows:

1. Verify that you have all the drawings in the set, and the specification book. Also verify that these are the most current documents.
2. Study the plot plan to understand the setting of the building.
3. Study the architectural pages to understand the layout of the structure. Look especially for offset and unusual levels. Also look carefully at systems or objects that extend beyond a single floor.
4. Review the foundation plan.
5. Review the wall construction.
6. Study the plumbing, mechanical, and electrical sheets.
7. Review all notes on the plans.

HOW TO READ A SET OF BLUEPRINTS *(cont.)*

8. Review the specifications and compare them to the drawings. (Specifications normally have priority.)

TITLE BLOCK

It is standard practice to include a title block on every page of a set of blueprints. The title block shows the name of the project and the name of the page. It usually shows the name of the architect, engineer, or designer as well.

Title blocks also show the date the drawing was made, and who made it. This information may be very important, because using an outdated set of drawings can cause serious problems.

OFFICE BUILDING FOR JAMES INDUSTRIES		
1628 Old Wine Rd.		Grayson & Mill Architects
North Platte, Nebraska		16 W. Hill St.
A3	Date: 4/18/05	
	By: JDE	North Platte, NE
	Ver: 1.2.0	State Reg. # 521930

PAGE NAMES AND NUMBERS

The pages in a set of blueprints are usually carefully lettered and numbered. The letters shown here are the most commonly used:

PAGE NAMES AND NUMBERS *(cont.)*

A—Architectural pages
S—Structural pages
P—Plumbing pages
M—Mechanical pages
E—Electrical pages

For example, a set of blueprints may consist of 24 pages, numbered as follows:

A1 through A4 (4 architectural pages)
S1 through S8 (8 structural pages)
P1 through P3 (3 plumbing pages)
M1 through M4 (4 mechanical pages)
E1 through E5 (5 electrical pages)

Architectural pages include not only plans for the building but plans for the surrounding areas as well.

Structural plans show the structure of the building: concrete, masonry, wood framing, roof plans, and so on.

Plumbing plans, obviously, show the plumbing system.

Mechanical plans show the heating, air conditioning, and process piping systems. They may also show systems such as conveyor lines.

Electrical plans show the building's electrical system.

OVERHEAD VIEW

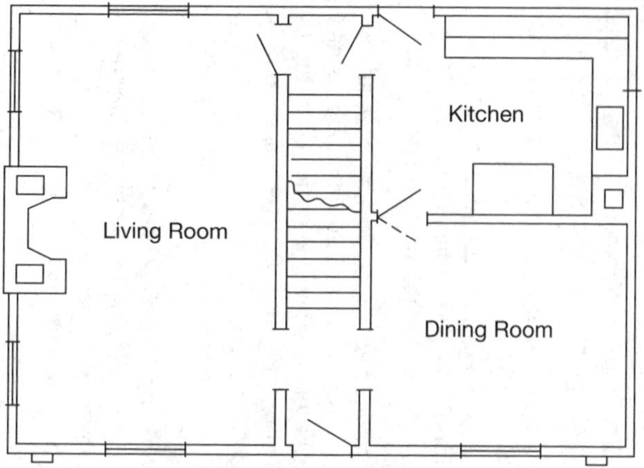

Kitchen

Dining Room

Living Room

A Typical First-Floor Plan of a House

This view would be available in real life by removing the roof and the entire second floor.

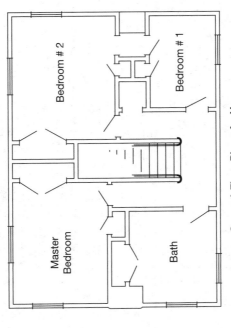

Second-Floor Plan of a House

This view would be available in real life if the roof were removed.

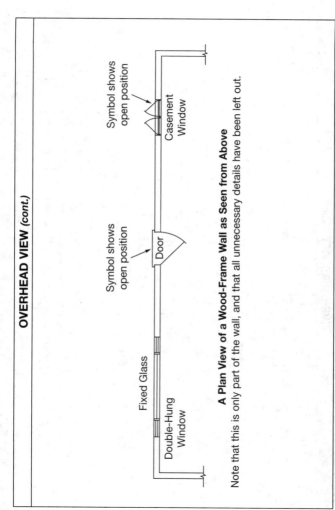

A Plan View of a Wood-Frame Wall as Seen from Above

Note that this is only part of the wall, and that all unnecessary details have been left out.

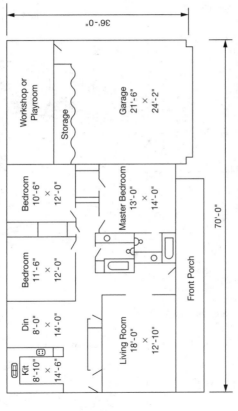

A one-floor home seen from above, as if the roof were removed. This drawing also specifies room and overall dimensions.

36'-0"

70'-0"

Workshop or Playroom

Storage

Garage
21'-6"
×
24'-2"

Bedroom
10'-6"
×
12'-0"

Master Bedroom
13'-0"
×
14'-0"

Bedroom
11'-6"
×
12'-0"

Din
8'-0"
×
14'-0"

Kit
8'-10"
×
14'-6"

Living Room
18'-0"
×
12'-10"

Front Porch

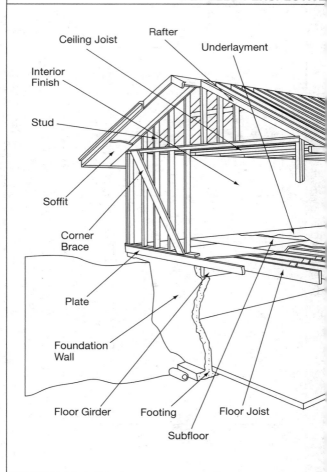

Rafter

Ceiling Joist

Underlayment

Interior Finish

Stud

Soffit

Corner Brace

Plate

Foundation Wall

Floor Girder

Footing

Floor Joist

Subfloor

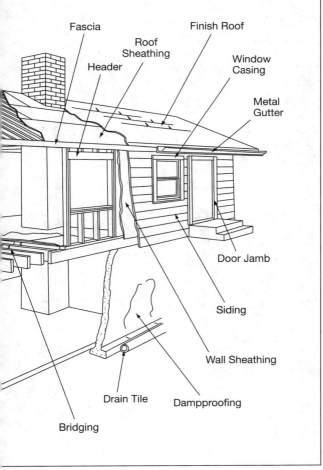

Fascia

Finish Roof

Roof Sheathing

Window Casing

Header

Metal Gutter

Door Jamb

Siding

Wall Sheathing

Drain Tile

Dampproofing

Bridging

SECTIONAL VIEW DEVELOPMENT

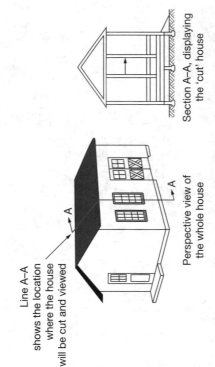

Line A–A
shows the location
where the house
will be cut and viewed

Perspective view of
the whole house

Section A–A, displaying
the 'cut' house

The drawing on the right is called an *elevation*. In this drawing we can see the house from top to bottom, viewing not from above, but from the side.

SCALE

Drawings are prepared according to a specific scale. For example, a very common scale for house construction is ¼" = 1'. On a drawing at this scale, a line 2 inches long represents a length of 8 feet.

For commercial buildings, a scale of ⅛" = 1' is more common. In this case a line 2 inches long represents a length of 16 feet in actual construction. The scale of a drawing should be shown on every page, generally at the bottom or side.

DRAWING SCALES

Plan Use	Ratio	English Units	Metric Units
Details:	1:1	12" = 1' 0" (full scale)	1000 mm = 1 m
	1:5	3" = 1' 0"	200 mm = 1 m
	1:10	1½" = 1' 0"	100 mm = 1 m
	1:20	½" = 1' 0"	50 mm = 1 m
Floor plans:	1:40	⅜" = 1' 0"	25 mm = 1 m
	1:50	¼" = 1' 0"	20 mm = 1 m
Plot plans:	1:80	3⁄16" = 1' 0"	13.3 mm = 1 m
	1:100	⅛" = 1' 0"	12.5 mm = 1 m
	1:200	1" = 20' 0"	5 mm = 1 m
Plat plans:	1:500	1" = 50' 0"	2 mm = 1 m
City maps (and larger):	1:1250	1" = 125' 0"	0.8 mm = 1 m
	1:2500	1" = 250' 0"	0.4 mm = 1 m

RULERS

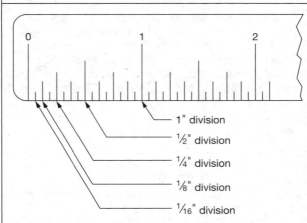

1" division
½" division
¼" division
⅛" division
1/16" division

The subdivisions on a full-size scale graduated in the English system of measurement.

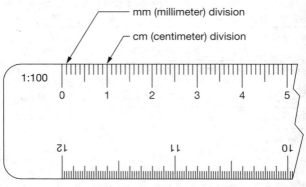

mm (millimeter) division
cm (centimeter) division

1:100

The subdivisions of a full-size scale graduated in the metric system of measurement.

1-14

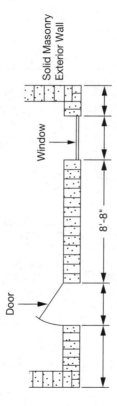

The arrows specify the distances for which dimensions are given. The large section of wall is shown as being 8 feet, 8 inches long.

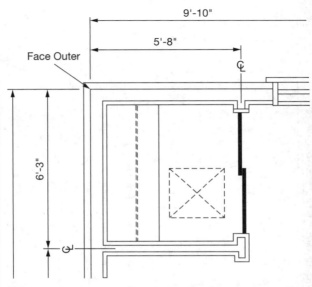

Dimensions showing lengths of interior and exterior walls.

Notation of floor joist size, span size and span direction.

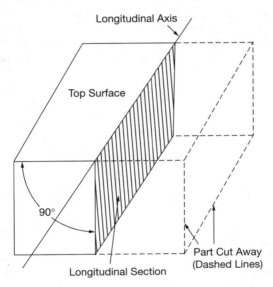

Longitudinal Axis

Top Surface

90°

Longitudinal Section

Part Cut Away
(Dashed Lines)

This is a pictorial view illustrating a longitudinal section, or a section across the length.

DIAGRAMS

In addition to bird's-eye-view plans, other diagrams can be used. Parts of a building of objects in the building can be shown from any angle—front, side, top, bottom, or anywhere in between.

It is necessary to use some imagination when viewing diagrams. You have to see the drawing and associate it with the object it represents in real life. If a drawing is properly done, this is not difficult. It does, however, require you to spend some time and effort.

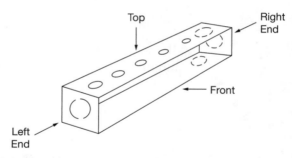

Pictorial drawing of a wire trough with top, front, and ends labeled.

DIAGRAMS (cont.)

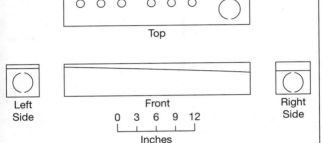

Top, front, and side views of the wire trough (shown on previous page).

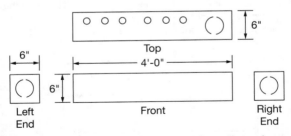

Same drawing as above showing various dimensions.

DIAGRAMS (cont.)

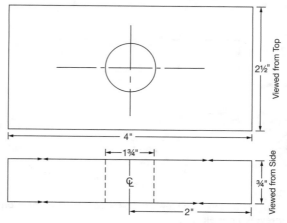

The above depicts a wood board with a round hole, and shows the top and side views. The dimensions are shown, the center line (ℂ) is shown, and the hole in the side view is shown with dashed lines. Note that the hole would not be visible from the side. Dashed lines indicate an edge that is hidden from view, and are used here to show the outline of the hidden hole.

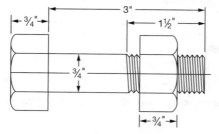

Dimensions of a bolt and nut.

SYMBOLS

Since it may be difficult or impossible to show every small item on a blueprint, we use symbols to represent many objects or groupings of objects. (A fairly small object, shown at small scale, might appear as a dot on a blueprint.) For example, electrical outlets are shown with symbols, as are switches, thermostats, heating vents, and so on.

A list of symbols should be shown on the plans, and usually there are several symbol lists. These symbols almost always conform to construction industry standards. Even if you are very familiar with industry standards, however, you should check the symbol list on the plans.

It is very important to pay attention to the symbols on a drawing. If you overlook a symbol, you could be overlooking an important part of the job, causing a lot of extra work later on in the process. You must always read a set of prints carefully and slowly, looking especially for the symbols that matter to you and to your work.

Remember that even the lines used on blueprints may represent specific symbols. In fact, various types of dashed and solid lines may be shown on symbol lists and noted as to what they represent. These may be used to show property lines, hidden lines, dimension lines, and others.

SYMBOLS (cont.)

	Earth
	Common or Face Brick
	Firebrick
	Concrete, General
	Lightweight Concrete
	Structural Concrete
	Concrete Block
	Cut Stone
	Rubble
	Cast Stone (Concrete)
	Wood Siding
	Wood Panel
	Wood Stud

	Rough Members
	Finished Members
	Plywood
	Wood Stud, Lath, and Plaster
	Metal Lath and Plaster
	Solid Plaster
	Lath and Plaster
	Shingles
	Glass
	Glass Block
	Glass
	Glass Block

MATERIAL SYMBOLS

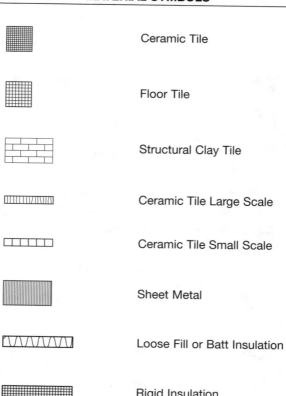

	Ceramic Tile
	Floor Tile
	Structural Clay Tile
	Ceramic Tile Large Scale
	Ceramic Tile Small Scale
	Sheet Metal
	Loose Fill or Batt Insulation
	Rigid Insulation
	Spray Foam Insulation

MATERIAL SYMBOLS (cont.)

Metals

Steel

Cast Iron

Bronz or Brass

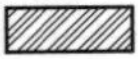

Aluminum

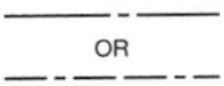

OR Structural Steel

LI Small Scale

L̗I Large Scale

● ◉ Rebars
L-Angles, S-Beams, etc.

MISCELLANEOUS SYMBOLS

North	N	Walk	═══════
Point of Beginning (POB)	◓	Improved Road	───────
Utility Meter or Valve	▲	Unimproved Road	-------
Power Pole and Guy	●→	Building Line	฿L
Light Standard	☼	Property Line	℗L
Traffic Signal	▯○	Property Line	─·─·─
Street Sign	─○─	Township Line	-------
Fire Hydrant	●├	Electric Service	──E── OR ······•······
Mailbox	⊠	Natural Gas Line	──G── OR ───────
Manhole	○	Water Line	──W── OR ·············
Tree	⊕	Telephone Line	──T── OR ----•----
Bush	○	Natural Grade	─ ─ ─ ─
Hedge Row	⋙⋙⋙	Finish Grade	───────
Fence	─·─·─	Existing Elevation	+XX.00'

1-26

WINDOW AND DOOR SYMBOLS

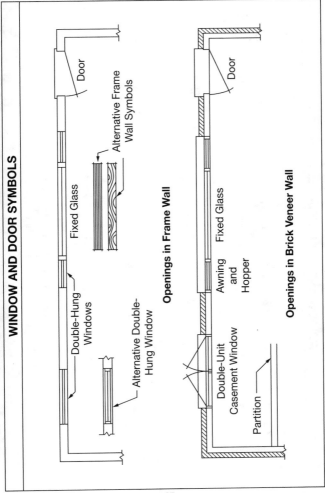

Door

Fixed Glass

Double-Hung Windows

Alternative Frame Wall Symbols

Alternative Double-Hung Window

Openings in Frame Wall

Door

Fixed Glass

Awning and Hopper

Double-Unit Casement Window

Partition

Openings in Brick Veneer Wall

1-27

WINDOW AND DOOR SYMBOLS *(cont.)*

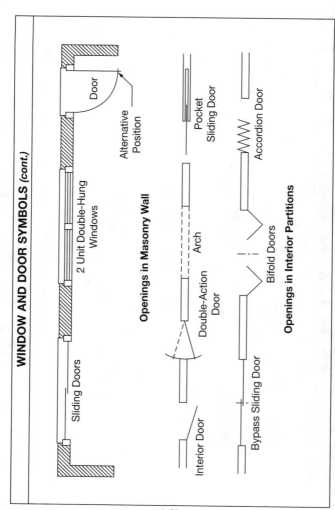

Sliding Doors

2 Unit Double-Hung Windows

Door

Alternative Position

Openings in Masonry Wall

Interior Door

Double-Action Door

Arch

Pocket Sliding Door

Bypass Sliding Door

Bifold Doors

Accordion Door

Openings in Interior Partitions

LINES USED IN BLUEPRINT DRAWINGS

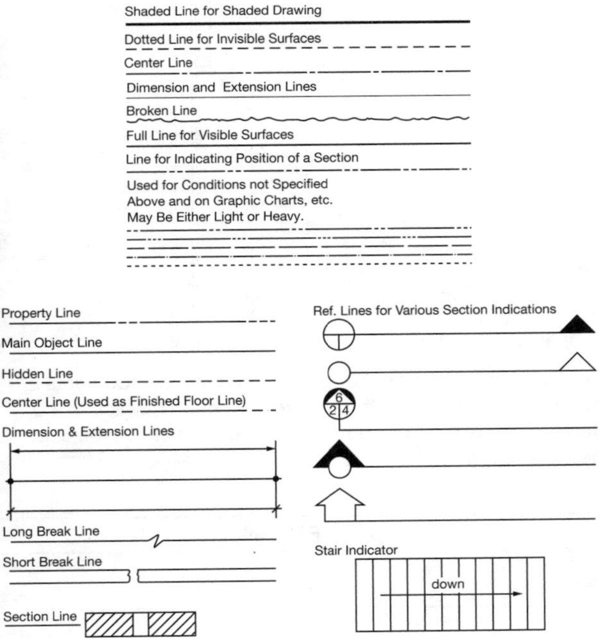

Shaded Line for Shaded Drawing

Dotted Line for Invisible Surfaces _ _ _ _ _ _ _ _

Center Line

Dimension and Extension Lines

Broken Line

Full Line for Visible Surfaces

Line for Indicating Position of a Section

Used for Conditions not Specified
Above and on Graphic Charts, etc.
May Be Either Light or Heavy.

Property Line

Main Object Line

Hidden Line _ _ _ _ _ _ _ _

Center Line (Used as Finished Floor Line) _ _

Dimension & Extension Lines

Long Break Line

Short Break Line

Section Line

Ref. Lines for Various Section Indications

Stair Indicator

down

HIDDEN LINE IN OVERHEAD VIEW

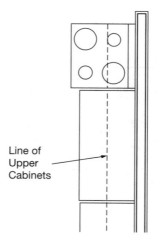

Line of
Upper
Cabinets

The dashed line shows the edge of the cabinets. Cabinets are
not shown, as they would obscure the counter, which is the
subject of this drawing. But since the placement of the cabinets is important, this edge is shown.

CENTER LINES

C/L

or

₵

Two methods of illustrating a center line.

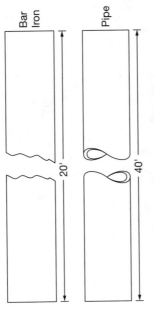

A Depiction of broken runs of bar and pipe.
Note that bars and pipes are shown differently, for clarity.

TOPOGRAPHIC LINES

Topographic lines are used to show the elevations of land above a reference, usually sea level. This drawing is of a hill, viewed from above. The base of the hill is at approximately 1100 feet above sea level. The peak of the hill rises above 1300 feet. The hill, from base to peak, is more than 200 feet high. We do not know, from the information given above, the width of the hill. If a scale were shown, that distance could be determined.

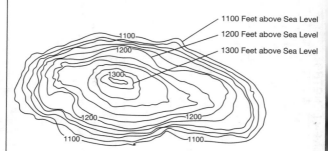

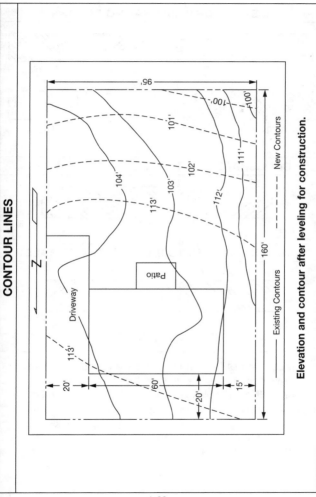

CONTOUR LINES

Elevation and contour after leveling for construction.

— — — New Contours

——— Existing Contours

95'

160'

100'

101'

102'

103'

104'

111'

112'

113'

113'

Patio

Driveway

N

20'

60'

20'

15'

1-33

METHOD OF ILLUSTRATING COLORS

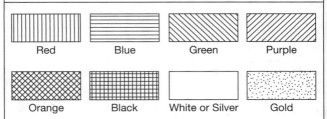

Red	Blue	Green	Purple
Orange	Black	White or Silver	Gold

CROSSHATCHING TO IDENTIFY TWO PIECES

CROSSHATCHING TO IDENTIFY TWO METALS

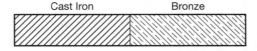

Cast Iron   Bronze

DETAILS

Most sets of blueprints include a number of detail drawings. These are used to illustrate a particularly difficult or complex portion of the building project.

In addition to appearing on a set of plans, the location of details are also shown. A special symbol is used to point out where the detail appears on the plan. The detail also shows this symbol next to it.

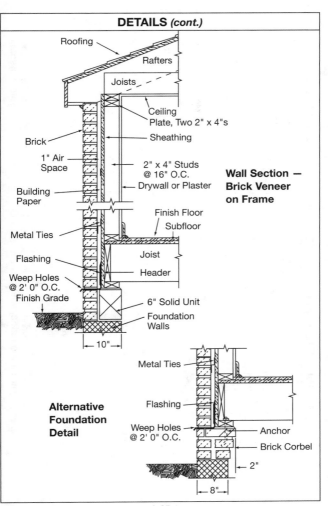

Wall Section — Brick Veneer on Frame

- Roofing
- Rafters
- Joists
- Ceiling Plate, Two 2" x 4"s
- Brick
- Sheathing
- 1" Air Space
- 2" x 4" Studs @ 16" O.C.
- Drywall or Plaster
- Building Paper
- Finish Floor
- Subfloor
- Metal Ties
- Joist
- Flashing
- Header
- Weep Holes @ 2' 0" O.C.
- Finish Grade
- 6" Solid Unit
- Foundation Walls
- 10"

Alternative Foundation Detail

- Metal Ties
- Flashing
- Weep Holes @ 2' 0" O.C.
- Anchor
- Brick Corbel
- 2"
- 8"

ANCHORING SILLS AND PLATES TO CONCRETE BLOCK WALLS

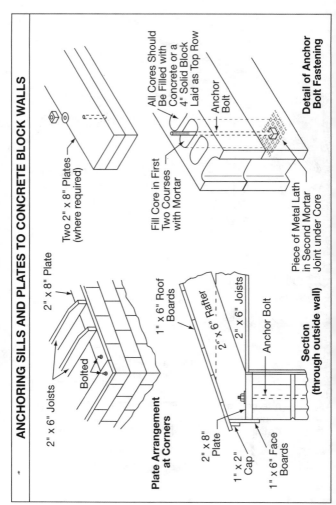

Two 2" x 8" Plates (where required)

All Cores Should Be Filled with Concrete or a 4" Solid Block Laid as Top Row

Fill Core in First Two Courses with Mortar

Anchor Bolt

Piece of Metal Lath in Second Mortar Joint under Core

Detail of Anchor Bolt Fastening

2" x 8" Plate

2" x 6" Joists

Bolted

Plate Arrangement at Corners

1" x 6" Roof Boards

2" x 6" Rafter

2" x 6" Joists

Anchor Bolt

Section (through outside wall)

2" x 8" Plate

1" x 2" Cap

1" x 6" Face Boards

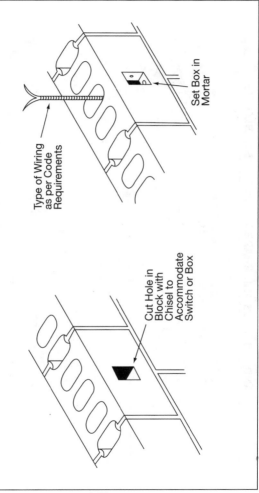

Type of Wiring
as per Code
Requirements

Set Box in
Mortar

Cut Hole in
Block with
Chisel to
Accommodate
Switch or Box

STANDARD SIZES AND SHAPES OF CONCRETE BLOCKS

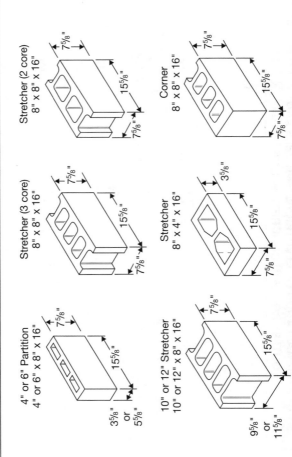

Stretcher (2 core)
8" x 8" x 16"

Corner
8" x 8" x 16"

Stretcher (3 core)
8" x 8" x 16"

Stretcher
8" x 4" x 16"

4" or 6" Partition
4" or 6" x 8" x 16"

10" or 12" Stretcher
10" or 12" x 8" x 16"

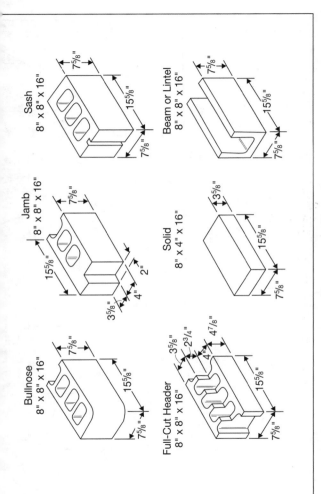

Sash
8" x 8" x 16"

Beam or Lintel
8" x 8" x 16"

Jamb
8" x 8" x 16"

Solid
8" x 4" x 16"

Bullnose
8" x 8" x 16"

Full-Cut Header
8" x 8" x 16"

12
1

Open above Divider Walls in
Dining, Kitchen, and Hall

Horizontal Decorative Beams

Top of Divider Halls

Fill Height Entry
Halls Beyond

2'-0"

Dining Room

Elevation 2'-0"

$\frac{6}{A5}$

$\frac{5}{A5}$

Basement (Utilities)

Elevation 7'-3"

Section $\frac{8}{A2}$

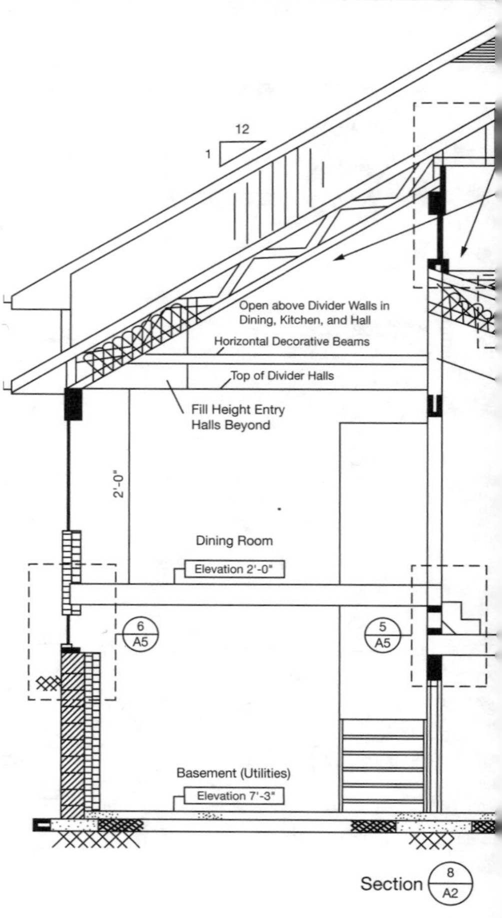

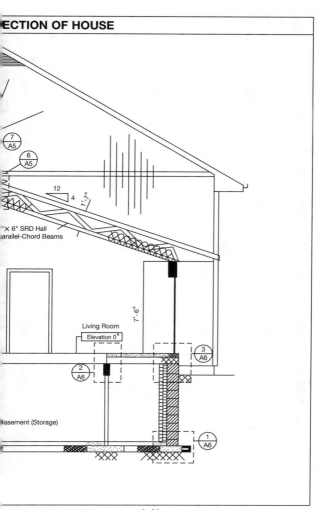

7
A5

6
A5

12
4

1'-7"

"× 6" SRD Hall
arallel-Chord Beams

7'-6"

Living Room

Elevation 0"

3
A6

2
A6

asement (Storage)

1
A6

CROSS-SECTION END VIEW OF A SIMPLE BLOCK WALL

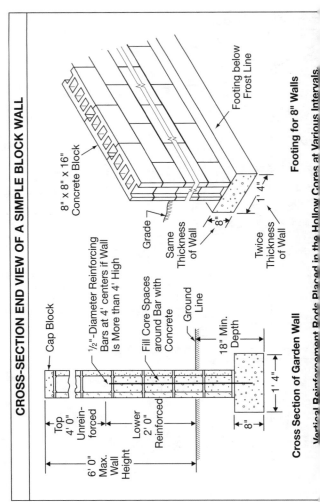

Footing for 8" Walls

Cross Section of Garden Wall

Vertical Reinforcement Rods Placed in the Hollow Cores at Various Intervals.

SCHEDULES

A schedule is a table made up of rows and columns, showing specific information. It appears on the pages where the things in the schedule are shown. For example, a door schedule will appear on the architectural pages, and will specify the dimensions and materials that are to be used for each door.

There are many types of schedules. Some of them include door schedules, window schedules, heating schedules, plumbing fixture schedules, motor schedules, and lighting fixture schedules.

DOOR SCHEDULE

Mark	Size	Amount Required	Remarks
1	2'0" × 6'8" × 1⅜"	3	Flush Door
2	2'6" × 6'8" × 1⅜"	4	Flush Door
3	2'6" × 6'8" × 1⅜"	1	Ext flush Door, 1 Light
4	3'0" × 7'0" × 1¾"	1	Ext flush Door, 4 Lights
5	1'8" × 6'8" × 1⅜"	1	Flush Door

WINDOW SCHEDULE

Mark	Size	Amount Required	Remarks
A	4'5⅛" × 4'2⅝"	3	Metal Frame
B	3'1⅛⅝" × 4'2⅝"	2	Metal Frame
C	3'1" × 4'2⅝"	1	Metal Frame
D	3'1" × 4'2⅝"	1	Metal Frame
E	1'7⅝" × 4'2⅝"	2	Metal Frame

FINISH SCHEDULE

Room	Floor	Walls	Ceiling	Baseboard	Trim
Dining and living	1" × 3" oak	½" Drywall Paint White	½" Drywall Paint White	Wood	Wood
Bedroom	1" × 3" oak	½" Drywall Paint White	½" Drywall Paint White	Wood	Wood
Bathroom	Linoleum-tan	½" Drywall Paint White	½" Drywall Paint White	Lino-cove	Wood
Kitchen	Linoleum-tan	½" Drywall Paint White	½" Drywall Paint White	Lino-cove	Wood
Utility Room	Linoleum-tan	½" Drywall Paint White	½" Drywall Paint White	Lino-cove	Wood
Hall	1" × 3" oak	½" Drywall Paint White	½" Drywall Paint White	Wood	Wood

1-44

ROOM FINISH SCHEDULE

No.	Name	Floor	Base	Walls				Ceiling		Remarks
				North	East	South	West	Material	Height	
101	Vestibule	Quarry Tile	Wood	Plaster Painted	Plaster Painted	Plaster Painted	Plaster Painted	"V" Groove	Varies	Provide Non Skid FL. Mat
102	Lounge	Q.T.	Wood	Plaster Painted	Plaster Painted Wainscot	Plaster Painted Wainscot	Plaster Painted	"V" Groove	Varies	
103	Bar	Q.T.	4" Vinyl	Plaster Painted	—	—	Plaster Painted	"V" Groove	9'-8"	
104	Dining Room	Slate	Wood	Wainscot	P.PTD/ Wains	P.PTD/ Wains	P.PTD/ Wains	Plaster PTD.	9'-8"	
105	Service	Q.T.	WD/Vinyl	P.PTD.	P.PTD.	—	P.PTD.	Plaster PTD.	9'-8"	
106	Corridor	Q.T.	Wood	P.PTD.	P.PTD.	—	P.PTD.	Plaster PTD.	7'-8"	
107	Women	C.T.	C.T.	P.PTD.	P.PTD.	P.PTD.	P.PTD.	Plaster PTD.	7'-8"	
108	Men	C.T.	C.T.	P.PTD.	P.PTD.	P.PTD.	P.PTD.	Plaster PTD.	7'-8"	
109	Kitchen	Conc.PTD	Vinyl	ST.STL/ C.T.	C.T./PTD. GWB	C.T./PTD. GWB	PTD. GWB	2×4-YWYL A.C.T	9'-8"	
110	Prep./Storage	Conc.PTD	Vinyl	PTD. GWB	PTD. GWB	—	PTD GWB	2×4-YWYL A.C.T	9'-8"	No Cls Above Walk-In

KITCHEN EQUIPMENT SCHEDULE

Equip. No.	Designation	H.P. or K.W.	Connection Volts	Connection Wire	Connection Conduit	Connection Prot.	Furnished by	Remarks
1	Freezer	1 H.P.	240 V	No. 12	3/4"	20 A	Equip by Others Outlet by Cont	24' A.F.F.; 1 φ
2	Mixer	1/5 H.P.	120 V	No. 12	3/4"	20 A		48" A.F.F.; 1 φ
3	Ice Machine	1 H.P.	120 V	No. 12	3/4"	20 A		72" A.F.F.; 1 φ
4	Refrigerated Display Case	1/4 H.P.	120 V	No. 12	3/4"	20 A		24" A.F.F.; 1 φ
5	Receptacle		120 V	No. 12	3/4"	20 A		72" A.F.F.; 1 φ
6	Refrigerator		120 V	No. 12	3/4"	20 A		6" A.F.F.; 1 φ
7	Ice Storage Chest		120 V	No. 12	3/4"	20 A		39" A.F.F.; 1 φ
8	Coffee Maker		240 V	No. 12	3/4"	20 A		39" A.F.F.; 3 φ
9	Milk Dispensers		120 V	No. 12	3/4"	20 A		39" A.F.F.; 1 φ
10	Waffle Maker		240 V	No. 10	3/4"	30 A		39" A.F.F.; 3 φ
11	Waffle Maker		240 V	No. 10	3/4"	30 A		39" A.F.F.; 3 φ
12	Refrigerator	1/4 H.P.	120 V	No. 12	3/4"	20 A		24" A.F.F.; 1 φ
13	Toaster		240 V	No. 12	3/4"	20 A		39" A.F.F.; 3 φ
14	Receptacle		120 V	No. 12	3/4"	20 A	→	72" A.F.F.; 1 φ

KITCHEN EQUIPMENT SCHEDULE *(cont.)*

Equip. No.	Designation	H.P. or K.W.	Connection				Furnished by	Remarks
			Volts	Wire	Conduit	Prot.		
15	Mixer		120 V	No. 12	¾"	20 A		6" A.F.F.; 1 φ
16	Updraft Unit	33.8	240 V	No. 1	1½"	125 A		26" A.F.F.; 3 φ
17	Upright Refrigerator	1.5	120 V	No. 12	¾"	20 A		74" A.F.F.; 1 φ
18	Ice Storage Chest		120 V	No. 12	¾"	20 A		39" A.F.F.; 1 φ
19	Coffee Maker		240 V	No. 12	¾"	20 A		39" A.F.F.; 9 φ
20	Refrigerator		120 V	No. 12	¾"	20 A		6" A.F.F.; 1 φ
21	Duplex Receptacle		120 V	No. 12	¾"	20 A		72" A.F.F.; 1 φ
22	Refrigerated Display Case	¼ H.P.	120 V	No. 12	¾"	20 A		24" A.F.F.; 1 φ
23	Floor Receptacle		120 V	No. 12	¾"	20 A		Floor Outlet; 1 φ
24	Floor Receptacle		120 V	No. 12	¾"	20 A		Floor Outlet; 1 φ
25	Refrigerator		120 V	No. 12	¾"	20 A	▶	72" A.F.F.; 1 φ

LIGHTING PANEL SCHEDULE

LP1

Ckt	Pole	Brkr	Load	Phase	Load	Brkr	Pole	Ckt
1	1	20	1600	A	1600	20	1	2
3	1	20	1200	B	650	20	1	4
5	1	20	675	C	1620	20	1	6
7	1	20	1250	A	SPARE	20	1	8
9	1	20	1620	B	875	20	1	10
11	1	20	1620	C	540	20	1	12
13	1	20	840	A	1440	20	1	14
15	1	20	1440	B	1440	20	1	16
17	1	20	1440	C	900	20	1	18
19	1	20	720	A	900	20	1	20
21	1	20	720	B	N/A	30	2	22
23	1	20	900	C				24
25	2	20	Utility	A	Utility			26
27				B				28
29	2	20	½ HP	C	3HP	20	3	30
31				A				32
33	1	20	Spare	B	3 HP	20	3	34
35	1	20	Spare	C				36
37	1	20	Spare	A				38
39			Space Only					40
41			Space Only					42

CONNECTED LOAD: 36 KW   FEED: 4 No. 1/0 TWIN 2" IMC

DESIGN LOAD: 40 KW   125 AMP SERVICE

PANEL : 42 CKT. 225 AMP MAIN LUGS ONLY
          SURFACE MOUNTED 3 ϕ 4 WIRE.

CHAPTER 2
Specifications

SPECIFICATION BASICS

Specifications are the written rules governing construction projects. You may frequently hear specifications referred to as *specs*. They are usually associated with a set of blueprints, but not always. Specifications have to be followed just as much as a set of blueprints have to, and sometimes more so.

For small projects, specifications may be written on plan pages, but for most projects, they will comprise a separate book.

Specifications contain not only the rules for performing the work, but also all of the contractual information. That is, they describe how everyone will be paid and what their responsibilities are. Therefore, it is very important that you know how to read specifications. Whether you are responsible for the installation end or the business end, you must know the rules.

Bear in mind that specifications are not easy-to read material. Reading a set of specs requires time and attention. If your only concern is installation rules, you can study the sections of the specifications that deal with those questions and read no further. However, if you are responsible for estimating, supervision, or finance, you will have to read a lot more.

WHAT TO LOOK FOR IN THE SPECIFICATIONS

General Conditions

- Liquidated damages
- Payment terms
- Retainage
- Owner, general contractor
- Schedules
- Specifiers
- Inspectors
- Chains of command and responsibility
- Bonding requirements
- Insurance requirements
- Change-order requirements
- Project close-out
- Safety requirements
- Training requirements
- Special conditions

Sitework

- Ground preparation and/or testing
- Trenching requirements and hazards
- Hauling and dumping
- Demolition
- Preexisting conditions
- Material and installation requirements
- Work required, but not shown on the prints
- Work also associated with other sections of the specs
- Special codes and/or inspections
- Special equipment
- Temporary requirements

Concrete

- Material and installation requirements
- Work required, but not shown on the prints
- Testing
- Special equipment
- Work also associated with other sections of the specs
- Special codes and/or inspections

Masonry

- Material and installation requirements
- Work required, but not shown on the prints
- Work also associated with other sections of the specs
- Special codes and/or inspections

Metals

- Material and installation requirements
- Work required, but not shown on the prints
- Work also associated with other sections of the specs
- Special codes and/or inspections

Wood and Plastics

- Material and installation requirements
- Work required, but not shown on the prints
- Work also associated with other sections of the specs
- Special codes and/or inspections

Thermal and Moisture Protection

- Material and installation requirements
- Work required, but not shown on the prints
- Work also associated with other sections of the specs
- Special codes and/or inspections

Doors and Windows

- Material and installation requirements
- Work required, but not shown on the prints
- Work also associated with other sections of the specs
- Special codes and/or inspections

Finishes

- Material and installation requirements
- Work required, but not shown on the prints
- Work also associated with other sections of the specs
- Special codes and/or inspections

Specialties

- Material and installation requirements
- Work required, but not shown on the prints
- Work also associated with other sections of the specs
- Special codes and/or inspections
- Temporary requirements

Equipment

- Material and installation requirements

- Work required, but not shown on the prints
- Work also associated with other sections of the specs
- Special codes and/or inspections
- Temporary requirements

Furnishings

- Material and installation requirements
- Work required, but not shown on the prints
- Work also associated with other sections of the specs
- Special codes and/or inspections
- Temporary requirements

Special Construction

- Material and installation requirements
- Work required, but not shown on the prints
- Work also associated with other sections of the specs
- Special codes and/or inspections
- Temporary requirements

Conveying Systems

- Material and installation requirements
- Work required, but not shown on the prints
- Work also associated with other sections of the specs
- Special codes and/or inspections
- Temporary requirements
- Equipment supplied by others

WHAT TO LOOK FOR IN THE SPECIFICATIONS *(cont.)*

Mechanical

- Material and installation requirements
- Work required, but not shown on the prints
- Work also associated with other sections of the specs
- Special codes and/or inspections
- Temporary requirements
- Equipment supplied by others
- Specialty work required

Electrical

- Material and installation requirements
- Work required, but not shown on the prints
- Work also associated with other sections of the specs
- Special codes and/or inspections
- Temporary requirements
- Equipment supplied by others
- Specialty work required

SPECIFICATION SECTIONS

Conditions of the Contract

00010 Pre-Bid Information
00100 Instructions to Bidders
00200 Information Available to Bidders
00300 Bid Forms
00400 Supplements to Bid Forms
00500 Agreement Forms
00600 Bonds and Certificates
00700 General Conditions
00800 Supplementary Conditions
00900 Addenda

Division 1
General Requirements

01010 Summary of Work
01020 Allowances
01025 Measurement and Payment
01030 Alternatives/Alternatives
01035 Modification Procedures
01040 Coordination
01050 Field Engineering
01060 Regulatory Requirements
01070 Identification Systems
01090 References
01100 Special Project Procedures
01200 Project Meetings
01300 Submittals
01400 Quality Control
01500 Construction Facilities and Temporary Controls
01600 Material and Equipment
01650 Facility Startup/ Commissioning
01700 Contract Closeout
01800 Maintenance

Division 2
Sitework

02010 Subsurface Investigation
02050 Demolition
02100 Site Preparation
02140 Dewatering
02150 Shoring and Underpinning
02160 Excavation Support Systems
02170 Cofferdams
02200 Earthwork
02300 Tunneling
02350 Piles and Caissons
02450 Railroad Work
02480 Marine Work
02500 Paving and Surfacing
02600 Utility Piping Materials
02660 Water Distribution
02680 Fuel and Steam Distribution
02700 Sewerage and Drainage
02760 Restoration of Underground Pipe
02770 Ponds and Reservoirs
02780 Power and Communications
02800 Site Improvements
02900 Landscaping

Division 3
Concrete

03100 Concrete Formwork
03200 Concrete Reinforcement
03250 Concrete Accessories
03300 Cast-In-Place Concrete
03370 Concrete Curing
03400 Precast Concrete
03500 Cementitious Decks and Toppings

SPECIFICATION SECTIONS *(cont.)*

03600 Grout
03700 Concrete Restoration and Cleaning
03800 Mass Concrete

Division 4
Masonry

04100 Mortar and Masonry Grout
04150 Masonry Accessories
04200 Unit Masonry
04400 Stone
04500 Masonry Restoration and Cleaning
04550 Refractories
04600 Corrosion Resistant Masonry
04700 Simulated Masonry

Division 5
Metals

05010 Metal Materials
05030 Metal Coatings
05050 Metal Fastening
05100 Structural Metal Framing
05200 Metal Joists
05300 Metal Decking
05400 Cold Formed Metal Framing
05500 Metal Fabrications
05580 Sheet Metal Fabrications
05700 Ornamental Metal
05800 Expansion Control
05900 Hydraulic Structures

Division 6
Wood and Plastics

06050 Fasteners and Adhesives
06100 Rough Carpentry
06130 Heavy Timber Construction
06150 Wood and Metal Systems
06170 Prefabricated Structural Wood
06200 Finish Carpentry
06300 Wood Treatment
06400 Architectural Woodwork
06500 Structural Plastics
06600 Plastic Fabrications
06650 Solid Polymer Fabrications

Division 7
Thermal and Moisture Protection

07100 Waterproofing
07150 Dampproofing
07180 Water Repellents
07190 Vapor Retarders
07195 Air Barriers
07200 Insulation
07240 Exterior Insulation and Finish Systems
07250 Fireproofing
07270 Firestopping
07300 Shingles and Roofing Tiles
07400 Manufactured Roofing and Siding
07480 Exterior Wall Assemblies
07500 Membrane Roofing
07570 Traffic Coatings
07600 Flashing and Sheet Metal
07700 Roof Specialties and Accessories
07800 Skylights
07900 Joint Sealers

SPECIFICATION SECTIONS *(cont.)*

Division 8
Doors and Windows

08100 Metal Doors and Frames
08200 Wood and Plastic Doors
08250 Door Opening Assemblies
08300 Special Doors
08400 Entrances and Storefronts
08500 Metal Windows
08600 Wood and Plastic Windows
08650 Special Windows
08700 Hardware
08800 Glazing
08900 Glazed Curtain Walls

Division 9
Finishes

09100 Metal Support Systems
09200 Lath and Plaster
09250 Gypsum Board
09300 Tile
09400 Terrazzo
09450 Stone Facing
09500 Acoustical Treatment
09540 Special Wall Surfaces
09545 Special Ceiling Surfaces
09550 Wood Flooring
09600 Stone Flooring
09630 Unit Masonry Flooring
09650 Resilient Flooring
09680 Carpet
09700 Special Flooring
09780 Floor Treatment
09800 Special Coatings
09900 Painting
09950 Wall Coverings

Division 10
Specialties

10100 Visual Display Boards
10150 Compartments and Cubicles
10200 Louvers and Vents
10240 Grilles and Screens
10250 Service Wall Systems
10260 Wall and Corner Guards
10270 Access Flooring
10290 Pest Control
10300 Fireplaces and Stoves
10340 Manufactured Exterior Specialties
10350 Flagpoles
10400 Identifying Devices
10450 Pedestrian Control Devices
10500 Lockers
10520 Fire Protection Specialties
10530 Protective Covers
10550 Postal Specialties
10600 Partitions
10650 Operable Partitions
10670 Storage Shelving
10700 Exterior Protection Devices for Openings
10750 Telephone Specialties
10800 Toilet and Bath Accessories
10880 Scales
10900 Wardrobe and Closet Specialties

SPECIFICATION SECTIONS (cont.)

Division 11
Equipment

11010 Maintenance Equipment
11020 Security and Vault Equipment
11030 Teller and Service Equipment
11040 Ecclesiastical Equipment
11050 Library Equipment
11060 Theater and Stage Equipment
11070 Instrumental Equipment
11080 Registration Equipment
11090 Checkroom Equipment
11100 Mercantile Equipment
11110 Commercial Laundry and Dry Cleaning Equipment
11120 Vending Equipment
11130 Audio-Visual Equipment
11140 Vehicle Service Equipment
11150 Parking Control Equipment
11160 Loading Dock Equipment
11170 Solid Waste Handling Equipment
11190 Detention Equipment
11200 Water Supply and Treatment Equipment
11280 Hydraulic Gates and Valves
11300 Fluid Waste Treatment and Disposal Equipment
11400 Food Service Equipment
11450 Residential Equipment
11460 Unit Kitchens
11470 Darkroom Equipment
11480 Athletic, Recreational, and Therapeutic Equipment
11500 Industrial and Process Equipment

11600 Laboratory Equipment
11650 Planetarium Equipment
11660 Observatory Equipment
11680 Office Equipment
11700 Medical Equipment
11780 Mortuary Equipment
11850 Navigation Equipment
11870 Agricultural Equipment

Division 12
Furnishings

12050 Fabrics
12100 Artwork
12300 Manufactured Casework
12500 Window Treatment
12600 Furniture and Accessories
12670 Rugs and Mats
12700 Multiple Seating
12800 Interior Plants and Planters

Division 13
Special Construction

13010 Air Supported Structures
13020 Integrated Assemblies
13030 Special Purpose Rooms
13080 Sound, Vibration, and Seismic Control
13090 Radiation Protection
13100 Nuclear Reactors
13120 Pre-Engineered Structures
13150 Aquatic Facilities
13175 Ice Rinks
13180 Site Constructed Incinerators
13185 Kennels and Animal Shelters
13200 Liquid and Gas Storage Tanks

SPECIFICATION SECTIONS (cont.)

13220 Filter Underdrains and Media

13230 Digester Covers and Appurtenances

13240 Oxygenation Systems

13260 Sludge Conditioning Systems

13300 Utility Control Systems

13400 Industrial and Process Control Systems

13500 Recording Instrumentation

13550 Transportation Control Instrumentation

13600 Solar Energy Systems

13700 Wind Energy Systems

13750 Cogeneration Systems

13800 Building Automation Systems

13900 Fire Suppression and Supervisory Systems

13950 Special Security Construction

Division 14
Conveying Systems

14100 Dumbwaiters

14200 Elevators

14300 Escalators and Moving Walks

14400 Lifts

14500 Material Handling Systems

14600 Hoists and Cranes

14700 Turntables

14800 Scaffolding

14900 Transportation Systems

Division 15
Mechanical

15050 Basic Mechanical Materials and Methods

15250 Mechanical Insulation

15300 Fire Protection

15400 Plumbing

15500 Heating, Ventilating, and Air Conditioning

15550 Heat Generation

15650 Refrigeration

15750 Heat Transfer

15850 Air Handling

15880 Air Distribution

15950 Controls

15990 Testing, Adjusting, and Balancing

Division 16
Electrical

16050 Basic Electrical Materials and Methods

16200 Power Generation— Built-Up Systems

16300 Medium Voltage Distribution

16400 Service and Distribution

16500 Lighting

16600 Special Systems

16700 Communications

16850 Electric Resistance Heating

16900 Controls

16950 Testing

GENERAL CONSTRUCTION CHECKLIST

Excavation
- ☐ Backfilling
- ☐ Clearing the site
- ☐ Compacting
- ☐ Dump fee/Hauling
- ☐ Equipment rental
- ☐ Equipment transport
- ☐ Establishing new grades
- ☐ General excavation
- ☐ Pit excavation
- ☐ Pumping
- ☐ Relocating utilities
- ☐ Removing obstructions
- ☐ Shoring
- ☐ Stripping topsoil
- ☐ Trenching

Demolition
- ☐ Cabinet removal
- ☐ Ceiling finish removal
- ☐ Concrete cutting
- ☐ Debris box
- ☐ Door removal
- ☐ Dump fee/Hauling
- ☐ Dust partition
- ☐ Electrical removal
- ☐ Equipment rental
- ☐ Fixtures removal
- ☐ Flooring removal
- ☐ Framing removal
- ☐ Masonry removal
- ☐ Plumbing removal
- ☐ Roofing removal
- ☐ Salvage value allowance
- ☐ Siding removal
- ☐ Slab breaking
- ☐ Temporary weather protection
- ☐ Wall finish removal
- ☐ Window removal

Concrete
- ☐ Admixtures
- ☐ Anchors
- ☐ Apron
- ☐ Caps
- ☐ Cement
- ☐ Columns
- ☐ Crushed stone
- ☐ Curbs
- ☐ Curing
- ☐ Drainage
- ☐ Equipment rental
- ☐ Expansion joints
- ☐ Fill
- ☐ Finishing
- ☐ Floating
- ☐ Footings
- ☐ Foundations
- ☐ Grading
- ☐ Gutters
- ☐ Handling
- ☐ Mixing
- ☐ Piers
- ☐ Ready mix
- ☐ Sand
- ☐ Screeds
- ☐ Slabs
- ☐ Stairs
- ☐ Standby time
- ☐ Tamping
- ☐ Topping
- ☐ Vapor barrier
- ☐ Waterproofing

Forms
- ☐ Braces
- ☐ Caps
- ☐ Cleaning for reuse
- ☐ Columns
- ☐ Equipment rental

Forms *(cont.)*
- ☐ Footings
- ☐ Foundations
- ☐ Key joints
- ☐ Layout
- ☐ Nails
- ☐ Piers
- ☐ Salvage value
- ☐ Slab
- ☐ Stair
- ☐ Stakes
- ☐ Ties
- ☐ Walers
- ☐ Wall

Reinforcing
- ☐ Bars
- ☐ Mesh
- ☐ Handling and Placing
- ☐ Tying

Masonry
- ☐ Arches
- ☐ Backing
- ☐ Barbecues
- ☐ Cement
- ☐ Chimney
- ☐ Chimney cap
- ☐ Cleaning
- ☐ Clean-out doors
- ☐ Dampers
- ☐ Equipment rental
- ☐ Fireplace
- ☐ Fireplace form
- ☐ Flashing
- ☐ Flue
- ☐ Foundation
- ☐ Glass block
- ☐ Handling
- ☐ Hearths
- ☐ Laying

Masonry *(cont.)*
- ☐ Lime
- ☐ Lintels
- ☐ Mantels
- ☐ Marble
- ☐ Mixing
- ☐ Mortar
- ☐ Paving
- ☐ Piers
- ☐ Reinforcing
- ☐ Repair
- ☐ Repointing
- ☐ Sand
- ☐ Sandblasting
- ☐ Sills
- ☐ Steps
- ☐ Stonework
- ☐ Tile
- ☐ Veneer
- ☐ Vents
- ☐ Walls and Wall ties
- ☐ Waterproofing

Rough Carpentry
- ☐ Area walls
- ☐ Backing
- ☐ Beams
- ☐ Blocking
- ☐ Bracing
- ☐ Bridging
- ☐ Building paper
- ☐ Columns
- ☐ Cornice
- ☐ Cripples
- ☐ Door frames
- ☐ Dormers
- ☐ Entrance hoods
- ☐ Fascia
- ☐ Fences
- ☐ Flashing

GENERAL CONSTRUCTION CHECKLIST *(cont.)*

Rough Carpentry *(cont.)*
- ☐ Framing clips
- ☐ Furring
- ☐ Girders
- ☐ Gravel stop
- ☐ Grounds
- ☐ Half timber work
- ☐ Hangers
- ☐ Headers
- ☐ Hip jacks
- ☐ Insulation
- ☐ Jack rafters
- ☐ Joists, ceiling
- ☐ Joists, floor
- ☐ Ledgers
- ☐ Nails
- ☐ Outriggers
- ☐ Pier pads
- ☐ Plates
- ☐ Porches
- ☐ Posts
- ☐ Rafters
- ☐ Ribbons
- ☐ Ridges
- ☐ Roof edging
- ☐ Roof trusses
- ☐ Rough frames
- ☐ Rough layout
- ☐ Scaffolding
- ☐ Sheathing, roof
- ☐ Sheathing, wall
- ☐ Sills
- ☐ Sleepers
- ☐ Soffit
- ☐ Stairs
- ☐ Straps
- ☐ Strong backs
- ☐ Studs
- ☐ Subfloor

Rough Carpentry *(cont.)*
- ☐ Timber connectors
- ☐ Trimmers
- ☐ Valley flashing
- ☐ Valley jacks
- ☐ Vents
- ☐ Window frames

Finish Carpentry
- ☐ Baseboard
- ☐ Bath accessories
- ☐ Belt course
- ☐ Built-ins
- ☐ Cabinets
- ☐ Casings
- ☐ Caulking
- ☐ Ceiling tile
- ☐ Closet doors
- ☐ Closets
- ☐ Corner board
- ☐ Cornice
- ☐ Counter tops
- ☐ Cupolas
- ☐ Door chimes
- ☐ Door hardware and stop
- ☐ Door jambs
- ☐ Door trim
- ☐ Doors
- ☐ Drywall
- ☐ Entrances
- ☐ Fans
- ☐ Flooring
- ☐ Frames
- ☐ Garage doors
- ☐ Hardware
- ☐ Jambs
- ☐ Linen closets
- ☐ Locksets
- ☐ Louver vents
- ☐ Mail slot

Finish Carpentry *(cont.)*
- ☐ Mantels
- ☐ Medicine cabinets
- ☐ Mirrors
- ☐ Molding
- ☐ Nails
- ☐ Paneling
- ☐ Rake
- ☐ Range hood
- ☐ Risers
- ☐ Roofing
- ☐ Room dividers
- ☐ Sash
- ☐ Door/window screens
- ☐ Shelving
- ☐ Shutters
- ☐ Siding
- ☐ Sills
- ☐ Sliding doors
- ☐ Stairs
- ☐ Stops
- ☐ Storm doors
- ☐ Threshold
- ☐ Treads
- ☐ Trellis
- ☐ Trim
- ☐ Vents
- ☐ Wallboard
- ☐ Window trim
- ☐ Wardrobe closets
- ☐ Weatherstripping
- ☐ Windows

Flooring
- ☐ Adhesive
- ☐ Asphalt tile
- ☐ Carpet
- ☐ Cork tile
- ☐ Flagstone
- ☐ Hardwood

Flooring *(cont.)*
- ☐ Linoleum
- ☐ Marble
- ☐ Nails
- ☐ Pad
- ☐ Rubber tile
- ☐ Seamless vinyl
- ☐ Slate
- ☐ Tack strip
- ☐ Terrazzo
- ☐ Tile
- ☐ Vinyl tile
- ☐ Wood flooring

Plumbing
- ☐ Bathtubs
- ☐ Bar sink
- ☐ Couplings
- ☐ Dishwasher
- ☐ Drain lines
- ☐ Dryers
- ☐ Faucets
- ☐ Fittings
- ☐ Furnace hookup
- ☐ Garbage disposers
- ☐ Gas service lines
- ☐ Hanging brackets
- ☐ Hardware
- ☐ Laundry trays
- ☐ Lavatories
- ☐ Medicine cabinets
- ☐ Pipe
- ☐ Pumps
- ☐ Septic tank
- ☐ Service sinks
- ☐ Sewer lines
- ☐ Showers
- ☐ Sinks
- ☐ Stack extension
- ☐ Supply lines

GENERAL CONSTRUCTION CHECKLIST *(cont.)*

Plumbing *(cont.)*
- [ ] Tanks
- [ ] Valves
- [ ] Vanity cabinets
- [ ] Vent stacks
- [ ] Washers
- [ ] Waste lines
- [ ] Water closets
- [ ] Water heaters
- [ ] Water meter
- [ ] Water softeners
- [ ] Water tank
- [ ] Water tap

Heating
- [ ] Air conditioning
- [ ] Air return
- [ ] Baseboard
- [ ] Bathroom
- [ ] Blowers
- [ ] Collars
- [ ] Dampers
- [ ] Ducts
- [ ] Electric service
- [ ] Furnaces
- [ ] Gas lines
- [ ] Grilles
- [ ] Hot water
- [ ] Infrared
- [ ] Radiant cable
- [ ] Radiators
- [ ] Registers
- [ ] Relocation of system
- [ ] Thermostat
- [ ] Vents
- [ ] Wall units

Roofing
- [ ] Adhesive
- [ ] Asbestos
- [ ] Asphalt shingles

Roofing *(cont.)*
- [ ] Built-up
- [ ] Canvas
- [ ] Caulking
- [ ] Concrete
- [ ] Copper
- [ ] Corrugated
- [ ] Downspouts
- [ ] Felt
- [ ] Fiberglass shingles
- [ ] Flashing
- [ ] Gravel
- [ ] Gutters
- [ ] Gypsum
- [ ] Hip units
- [ ] Insulation
- [ ] Nails
- [ ] Ridge units
- [ ] Roll roofing
- [ ] Scaffolding
- [ ] Shakes
- [ ] Sheet metal
- [ ] Slate
- [ ] Tile
- [ ] Tin
- [ ] Vents
- [ ] Wood shingles

Sheet Metal
- [ ] Access doors
- [ ] Caulking
- [ ] Downspouts
- [ ] Ducts
- [ ] Flashing
- [ ] Gutters
- [ ] Laundry chutes
- [ ] Roof flashing
- [ ] Valley flashing
- [ ] Vents

GENERAL CONSTRUCTION CHECKLIST *(cont.)*

Electrical Work
- ☐ Air conditioning
- ☐ Appliance hook-up
- ☐ Bell wiring
- ☐ Cable
- ☐ Ceiling fixtures
- ☐ Circuit breakers
- ☐ Circuit load adequate
- ☐ Clock outlet
- ☐ Conduit
- ☐ Cover plates
- ☐ Dimmers
- ☐ Dishwashers
- ☐ Dryers
- ☐ Fans
- ☐ Fixtures
- ☐ Furnaces
- ☐ Garbage disposers
- ☐ High voltage line
- ☐ Hood hook-up
- ☐ Hook-up
- ☐ Lighting
- ☐ Meter boxes
- ☐ Ovens
- ☐ Panel boards
- ☐ Plug outlets
- ☐ Ranges
- ☐ Receptacles
- ☐ Relocation of existing lines
- ☐ Service entrance
- ☐ Switches
- ☐ Switching
- ☐ Telephone outlets
- ☐ Television wiring
- ☐ Thermostat wiring
- ☐ Transformers
- ☐ Vent fans
- ☐ Wall fixtures
- ☐ Washers

Electrical Work *(cont.)*
- ☐ Water heaters
- ☐ Wire

Plastering
- ☐ Bases
- ☐ Beads
- ☐ Cement
- ☐ Coloring
- ☐ Cornerite
- ☐ Coves
- ☐ Gypsum
- ☐ Keene's cement
- ☐ Lath
- ☐ Lime
- ☐ Partitions
- ☐ Sand
- ☐ Soffits

Painting and Decorating
- ☐ Aluminum paint
- ☐ Cabinets
- ☐ Caulking
- ☐ Ceramic tile
- ☐ Concrete
- ☐ Doors
- ☐ Draperies
- ☐ Filler
- ☐ Finishing
- ☐ Floors
- ☐ Masonry
- ☐ Paperhanging
- ☐ Paste
- ☐ Roof
- ☐ Sandblasting
- ☐ Shingle stain
- ☐ Stucco
- ☐ Wallpaper removal
- ☐ Windows
- ☐ Wood

GENERAL CONSTRUCTION CHECKLIST *(cont.)*

Glass and Glazing
- ☐ Breakage allowance
- ☐ Crystal
- ☐ Hackout
- ☐ Insulating glass
- ☐ Mirrors
- ☐ Obscure
- ☐ Ornamental
- ☐ Plate
- ☐ Putty
- ☐ Reglaze
- ☐ Window glass

Indirect Costs
- ☐ Barricades
- ☐ Bid bond
- ☐ Builder's risk insurance
- ☐ Building permit fee
- ☐ Business license
- ☐ Clean-up
- ☐ Completion bond
- ☐ Debris removal
- ☐ Design fee
- ☐ Equipment floater insurance
- ☐ Equipment rental
- ☐ Estimating fee
- ☐ Expendable tools
- ☐ Field supplies
- ☐ Job shanty and or utilities
- ☐ Job signs
- ☐ Liability insurance
- ☐ Maintenance bond
- ☐ Patching after subcontractors
- ☐ Payment bond
- ☐ Plan checking fee
- ☐ Plan cost
- ☐ Protecting adjoining property
- ☐ Protection during construction
- ☐ Removing utilities
- ☐ Repairing damage

Indirect Costs *(cont.)*
- ☐ Sales commission
- ☐ Sales taxes
- ☐ Sewer connection fee
- ☐ State contractor's license
- ☐ Street closing fee
- ☐ Street repair bond
- ☐ Supervision
- ☐ Survey
- ☐ Temporary electrical
- ☐ Temporary fencing
- ☐ Temporary heating
- ☐ Temporary lighting
- ☐ Temporary toilets
- ☐ Temporary water
- ☐ Transportation equipment
- ☐ Travel expense
- ☐ Water meter fee
- ☐ Waxing floors

Administrative Overhead
- ☐ Accounting
- ☐ Advertising
- ☐ Automobiles
- ☐ Depreciation
- ☐ Donations
- ☐ Dues and subscriptions
- ☐ Entertaining
- ☐ Interest
- ☐ Legal fees
- ☐ Licenses and fees
- ☐ Office insurance
- ☐ Office rent
- ☐ Office salaries/benefits
- ☐ Office utilities
- ☐ Postage
- ☐ Repairs
- ☐ Taxes
- ☐ Uncollectible accounts

CHAPTER 3
Site Plans

SITE PLAN SYMBOLS AND ABREVIATIONS

abut.	Abutment	°, deg.	Degree
acre	Acre	°C	Degree celsius
aggr.	Aggregate	DHV	Design hour volume
AH	Ahead	diag.	Diagonal
alt.	Alternate	dia. or D	Diameter
&	And	diaph.	Diaphragm
et al	And others	Dist.	District
et ux	And wife	DLC	Donation land claim
etc.	And so forth (et cetera)	drwg(s).	Drawing(s)
appr.	Approach		
approx.	Approximate	E	East
asph.	Asphalt	EP	Edge of pavement
ADT	Average daily traffic	EW	Edge of water
		ER	Edge of road
Bk	Back	elev.	Elevation
b. to b.	Back to back	El. 94,066	Elevation with number
BP	Balance point	emb.	Embankment
btr.	Batter	engr(s).	Engineer(s)
bm.	Beam	EQ or eq.	Equation
brg.	Bearing	exc.	Excavation
beg.	Beginning	exp. jt.	Expansion joint
BM	Bench mark		
br.	Bridge	Fed.	Federal
		fin.	Finish
℄	Centerline	flg.	Flange
cc or c. to c.	Center to center	ft	Foot
ctrs.	Centers	ftg.	Footing
ch. ch.	Channel change	e.g.	For example
clr.	Clear		
col.	Column	galv.	Galvanized
conc.	Concrete	ga.	Gage (gauge)
conn.	Connection		
constr.	Construction	hdwl.	Headwall
constr. jt.	Construction joint	hex.	Hexagon
CTSM	Contingent sum	HW	High water
cont.	Continuous	hwy.	Highway
corr.	Corrugated	hse.	House
CMP	Corrugated metal pipe	HES	Homestead entry survey
C	Coulomb		
co.	County	Iden.	Identification
ctsk.	Countersink	In.	Inch
cr.	Creek	Incl.	Inclusive
yd³ or yd3	Cubic yard	ID	Inside diameter
culv.	Culvert		
Δ	Curve delta	jt.	Joint
LT	Curve left	lam.	Lamination
RT	Curve right	lat.	Latitude
		lt.	Left

L	Length of curve	rad	Radian
L	Liter	R	Radius
long.	Longitudinal	R.	Range
LW	Low water	reconst.	Reconstruction
LPSM	Lump sum	reinf.	Reinforcement
		reqd.	Required
mag.	Magnetic	res.	Reservior or reservation
maint.	Maintenance	ret. wall	Retaining wall
matl.	Material	rt.	Right
max.	Maximum	R/W	Right-of-way
ml.	Mile	rd.	Road
mph	Miles per hour	rdwy.	Roadway
M.P.	Mile post	rte.	Route
'	Minute(s) (angular)	sch.	School
min.	Minimum	"	Second (angular)
misc.	Miscellaneous	s	Second (time)
mon.	Monument	sec.	Section
mtn(s).	Mountain(s)	sl. prot.	Slope protection
		S	South
neg.	Negative	spa.	Spacing, spaces or spaced
N	North	spec.	Specification
no.	Number	sq	Square
OG	Original ground	yd² or yd2	Square yard
o. to o.	Out to out	std.	Standard
OD	Outside diameter	sta.	Station
		stiff.	Stiffener
pvmt.	Pavement	str.	Straight
pct. or %	Percent	st.	Street
perf.	Perforate	stgr.	Stringer
pl.	Plate	struc.	Structural
PCC	Point of compound curve	e	Super elevation rate
PC	Point of curve	sym.	Symmetrical
POC	Point on curve	tan.	Tangent
PI	Point of intersection	T	Tangent length
PSC or SC	Point of spiral to curve	TBM	Temporary bench mark
PCS or CS	Point of curve to spiral	i.e.	That is
POS	Point on spiral	thd.	Thread
PST or ST	Point of spiral to tangent	T.	Township
POT	Point on tangent	typ.	Typical
PS or TS	Point of tangent to spiral		
PT	Point of tangent	vph	Vehicles per hour
lb	Pound	VPI	Vertical point of intersection
psi	Pressure (pounds/sq. inch)	whs.	Warehouse
proj.	Project	W	West
quant.	Quantities		

National boundary		
State boundary		
County boundary		
City boundary		
Township or range line		
Section line		
1/4 section line		
1/16 section line		
National park or forest boundary		
Property line		P/L
Right-of-way line	Existing	r/w
	Proposed	R/W
Right-of-way line with monument	Existing	r/w
	Proposed	R/W
Partial control of access	Existing	
	Proposed	
Full control of access	Existing	
	Proposed	
Easement (Permanent—temporary)	PE	TCE
Slope stake	Top of cut	
	Toe of fill	
Roadway, existing		
Railroad	Single track	
	Multiple track	
Trail		
Intermittent drainage and small creek		
Large creek		
River		
Lake, pond or reservoir, marshland		
Spring		
Treeline, tree		
Material source		

	FOUND	PROJECTED
Section corner	36⌄31 / 1▲6	36⌄31 / 1▲6
1/4 section corner	15 / 22	15 / 22
1/16 section or property corner	——●——	——○——
Property corner	●	No symbol
Parcel number	No symbol	(100)

SITE PLAN SYMBOLS AND ABREVIATIONS *(cont.)*

North arrow			

REG	STATE	PROJECT	SHEET NO.	TOTAL SHEETS

	EXISTING	PROPOSED
Fence	x ———— x	xx ———— xx
Gate with fence	—x⋈x—	—xx▶◀xx—
Cattleguard	⟨⊞⊞⊞⊞⟩	⟨⊞⊞⟩
Guardrail	••••••••••••	••••••••••••
Median & side barrier	••••••••••••	————————
Signs — Post mounted	▽	▼
Portable	No symbol	ᴛᴛ
Commercial	▽▽	▼▼
Retaining wall		Wall face
Power pole with utility line	••••••• P •••••••	——— P ———
Telephone pole with utility line	••••••• T •••••••	——— T ———
Joint use pole with utility lines	•••••• P&T ••••••	——— P&T ———
Support pole, and with anchor	o —o→	• —•→
Telephone booth or pedestal	□ TB or TP	■ TB or TP
Street light	—☆ or ☼	—★ or ✹
Underground utilities G=gas, O=oil, P=power, SA=sanitary sewer, SS=storm sewer, T=telephone, W=water	••••••⊣w⊢••••••	—⊣W⊢—
Bridge	⌐••••••••⌐	
Pipe culvert (arrow shows flow)	••••••••►	————►
Pipe culvert with end section	▷••••••••	▶————
Pipe culvert with headwall	⊢••••••••⊣	⊢———⊣
Box culvert	⌐••••••••⌐	
Culvert with drop inlet	□••••••••	■————
Underdrain	⊢•••⊣UD⊢•••⊣	⊢—⊣UD⊢—⊣
	T-45	
Traverse point (horizontal and vertical) top of triangle points north	▲ 645,934	No symbol
	T-3	
Traverse point (horizontal)	⊕	No symbol
Brass cap	▲	No symbol
Steel pin	●	No symbol
Hub & tack	○	No symbol
Spot elevation	×99.9	No symbol
Coordinate grid tick	+	No symbol
Building	⌐••••••⌐	⌐——⌐
Boring location	◕	No symbol
Riprap	No symbol	(riprap pattern)

3-4

LEGAL DESCRIPTIONS

Any construction site must be carefully identified. This is done by giving it a legal description. The legal description must be prepared carefully, so that it describes this piece of land, and no other piece of land.

The legal description must be accurate, with no possibility of confusion. In a city or built-up area, this is usually assured by reference to a lot and/or block within an existing subdivision. A common description of this type might say:

"Lot 5, Block 6, TOUHY ESTATES, Country of Clarke, State of South Dakota."

In a rural area, reference may be made to Sections, Townships, and Ranges, A description of this type might say:

"The northwest quarter of the southwest quarter of Section 10, Township 2 East, Range 41 West of the Fifth Principal Meridian."

The word *quarter* indicates the division of a one-square-mile section into quadrants (quarters), with each quarter being progressively divided into further quarters.

Legal descriptions (Which can be more complicated than the examples provided here), are shown on all site plans.

LEGAL DESCRIPTIONS *(cont.)*

R 67W OF 6TH P.M.

6	5	4	3	2	1
7	8	9	10	11	12
18	17	16	15	14	13
19	20	21	22	23	24
30	29	28	27	26	25
31	32	33	34	35	36

u 2 N

N

Illustrating a large plot of land called a township, which is divided into one-square-mile sections.

CONTOURS

A site plan will also show the contours of the land. These are shown with topographical drawings. Be aware that the site drawings may show the contours of the land as it exists prior to construction and how it will be changed during the construction process.

UTILITIES

Another important feature of site plans is the existence or requirement for utility lines.

If power, sewer, gas, or other utility lines exist on the property, they should be shown on the site drawings. Or, the designer may simply include a note saying that he or she does not know what utility lines may be present.

The site plan also shows which utility lines will be required. If you may be responsible for a power line, or a sewer line, telephone lines, or any other type of utility service, the site plans contain very important information for you. You must review them carefully.

PLOT PLAN

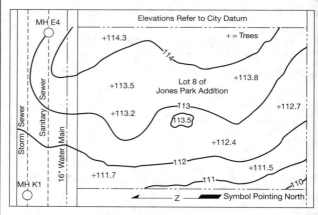

This plot plan shows the contours of the land on a construction site.

Notice the utility lines at the left. Also note that the elevations come from "city datum," which can be checked at the local city hall.

The "MH" notes associated with the round symbols refer to manholes. For example, the sanitary sewer manhole at the northeast of the site is designated manhole number E4.

CONTOUR LINES

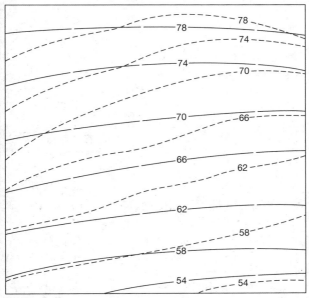

Existing Grade
Finished Grade

Contour lines showing the original and finished grade for a construction site.

SITE PLAN

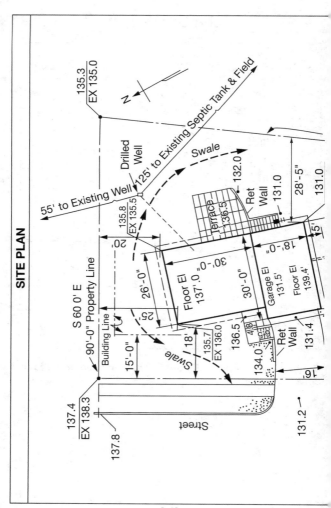

3-10

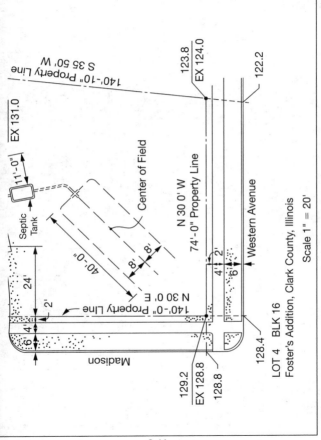

Center of Field

40'-0"

8'

8'

8'

N 30 0' E

140'-0" Property Line

N 30 0' W
74'-0" Property Line

S 35 50' W
140'-10" Property Line

EX 131.0

11'-0"

Septic
Tank

24'

2'

4'

6'

Madison

129.2

EX 128.8

128.8

2'

4'

6'

128.4

2'

123.8

EX 124.0

122.2

Western Avenue

LOT 4   BLK 16
Foster's Addition, Clark County, Illinois

Scale 1" = 20'

3-11

1P
3

R_L

Make New Pavement Flush
with Existing Pavement

137
136
135
134
133
132

124
35"

123

Existing Hedgerow along Ditch

114

134A

11.44

R=25'

132
131
131

New Sewage Lift Station
for Detail, See
type Drawing 683334

129

125

186.20   33.06'-QP
VP.HD.122
30'CW

10" Storm Sewer
for Detail, See
Field Drawing 68383

4

New Concrete
Road

2.25

IP

100'-30'E   200-26

81

Legend		
- - - - Existing		►── Drainage Flow
──── New		●── Utility Pole

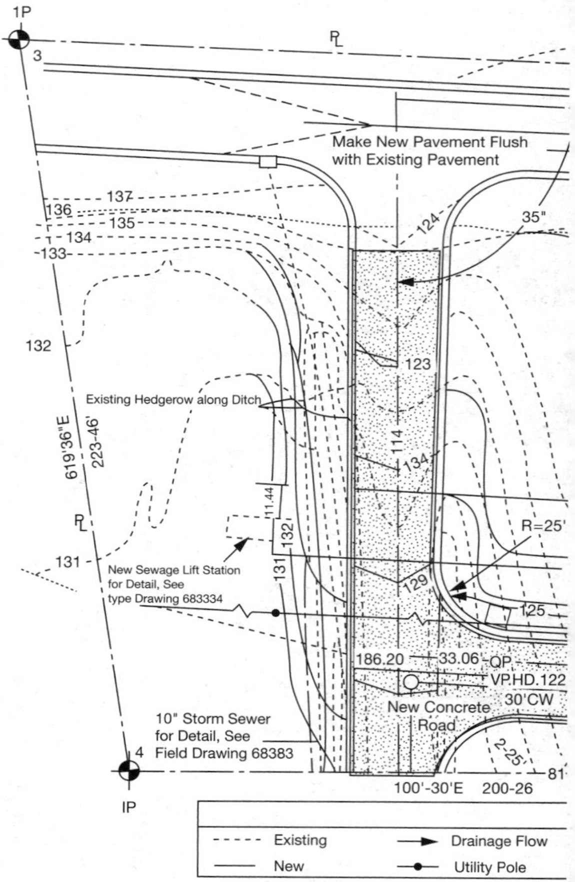

This is a complex site plan of the type

3-12

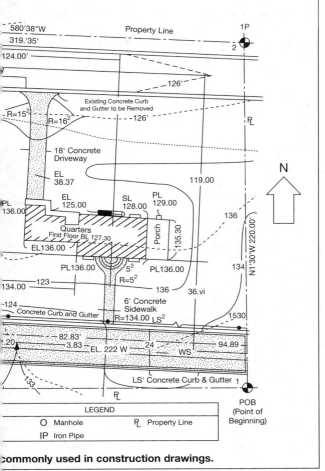

580'38"W Property Line 1P
319.'35' 2
124.00'

126'

Existing Concrete Curb
and Gutter to be Removed
R=15² R=16² 126' R̶L

18' Concrete
Driveway

EL
38.37 119.00

PL EL SL PL
136.00 125.00 128.00 129.00 136

Quarters Porch 135.30
First Floor BL 127.30
EL136.00

PL136.00 5² PL136.00 134
R=5²

134.00 123 136 36.vi

124 6' Concrete
Concrete Curb and Gutter Sidewalk
R=134.00 LS² 1530

82.83'
20 3.83 EL. 222 W 24 WS 94.89

133 LS' Concrete Curb & Gutter 1

R̶L POB
(Point of
Beginning)

N N1'30"W 220.00'

LEGEND	
O Manhole	R̶L Property Line
IP Iron Pipe	

commonly used in construction drawings.

3-13

TOPOGRAPHICAL SYMBOLS

![hard surface road]	Hard Surface Road
![unimproved road]	Unimproved Road
![railroad track]	Railroad Track
![power line]	Power Line
![telephone line]	Telephone Line, Pipeline (labeled as to type)
![property line]	Property Line
![sand]	Sand
![gravel]	Gravel
![water]	Water
![open woods]	Open Woods
![small trees]	Small, Irregularly Spaced Trees
![cultivated area]	Cultivated Area
![tall grass]	Tall Grass
![marsh]	Marsh
![large stones]	Large Stones

TOPOGRAPHICAL SYMBOLS *(cont.)*

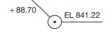 Spot Elevation

 Property Corner with Monument

BM × 868.75
BM △ 149.26   Bench Marks

Trees

Ground Cover

SYMBOLS FOR MATERIALS AND LAND

 Common Brick

 Concrete Block

 Elevation Concrete Block

 Face Brick

 Concrete Block

 Fire Brick

 Loose Fill or Batte Insulation

SYMBOLS FOR MATERIALS AND LAND (cont.)

	Board or Quilt Insulation
	Solid or Cork Insulation
	Stud, Lath and Plaster
	Solid Plaster Wall
	Plastered Wall
	I-Beam
	Angle Iron
	Reinforcing Iron Bars
	Water
	Puddle
	Concrete
	Brick

SYMBOLS FOR MATERIALS AND LAND *(cont.)*

	Coursed, Uncoursed Rubble
	Ashler
	Rock
	Sand
Material Name	Other Materials
	Swampland
	River
	Lake
† † † † † †	Cemetery
	School
	Church
	Wood
	Glass

SYMBOLS FOR MATERIALS AND LAND *(cont.)*

	Cast Iron
	Cast Steel
	Wrought Steel
	Wrought Iron
	Wrought Steel
	Babbitt, Lead or White Metal
	Wrought Iron
	Copper, Brass, or Composition
	Aluminum
	Rubber, Vulcanite or Insulation
	Property Line
	Center Line
	Building

LANDSCAPE SYSTEMS AND GRAPHICS

Window

Door

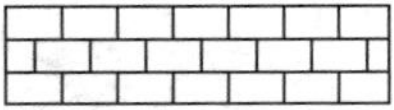

Paving

Pattern

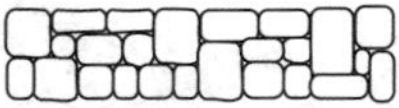

Random

Wall

Stone Wall

Hedge

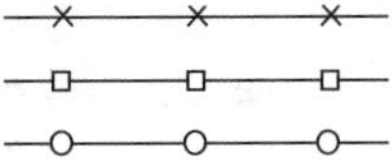

Fence

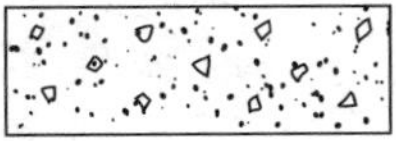

Concrete

Sand

LANDSCAPE SYSTEMS AND GRAPHICS (cont.)

Brick

Gravel

Rock

Water

Swamp

Slope

Steps

LANDSCAPE SYSTEMS AND GRAPHICS (cont.)

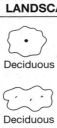

Trees

Deciduous   Evergreen

Shrubs

Deciduous   Evergreen

Herbaceous Plants (Flowers)

Same Variety

Grass

Ground Cover

Benchmark

 El.00.0

Topographic Contours

Contour Lines

—— —— —— ——

Unaltered

—— — —— — ——

Altered

——————————

Proposed

MATERIALS SYMBOLS

Earthworks

Earth/Compact Fill

Porous Fill/Graval

Rock

Concrete

Cast-in-Place/Precast

Lightweight

Sand/Mortar/Plaster/
Cut Stone

Masonry

Adobe/Rammed Earth

Common/Face

Firebrick

Concrete Block

MATERIALS SYMBOLS *(cont.)*

	Gypsum Block
	Structural Facing Tile

Stone

	Bluestone/Slate/ Soapstone/Flagging
	Rubble
	Marble

Metal

	Aluminum
	Brass/Bronze
	Steel/Other Metals

Wood

	Finish

MATERIALS SYMBOLS *(cont.)*

Rough

Blocking

Hardboard

Plywood—Large Scale

Plywood—Small Scale

CHAPTER 4
Architectural Drawings

ARCHITECTURAL DRAWINGS

Architectural drawings normally include elevation drawings of all the exterior surfaces of the building. They also include floor plans showing all walls, doors, and partitions.

Architectural pages also include the plans and details of foundations, walls, floors, ceilings, and roof construction. They also show all the floor levels, which is important to recognize in a complex structure.

BASEMENT PLAN

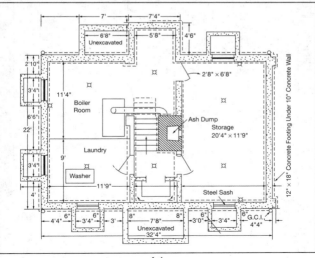

4-1

FOUNDATION PLAN

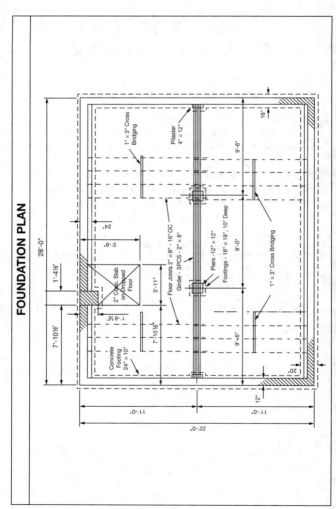

FOOTING PLAN

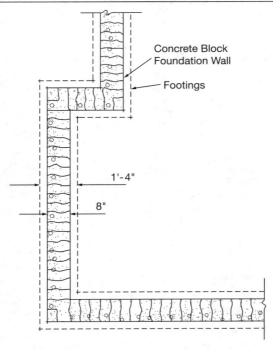

Concrete Block
Foundation Wall

Footings

1'-4"

8"

This overhead view shows a heavy foundation wall above the wider footings below. The dashed lines indicate that the footings are at a different level or below grade.

FOOTING AND WALL DETAIL

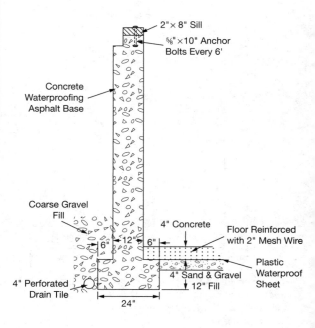

2" × 8" Sill

⅝" ×10" Anchor Bolts Every 6'

Concrete Waterproofing Asphalt Base

Coarse Gravel Fill

4" Concrete

Floor Reinforced with 2" Mesh Wire

Plastic Waterproof Sheet

6" 12" 6"

4" Perforated Drain Tile

4" Sand & Gravel 12" Fill

24"

4-4

FOOTING DETAIL

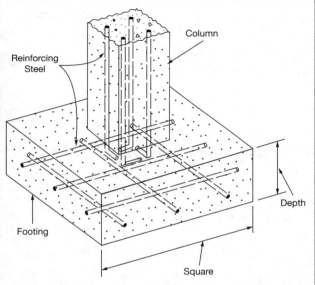

Column

Reinforcing
Steel

Depth

Footing

Square

This is a perspective drawing. Notice that the rebar is
shown inside of the footing and column. This is indicated
by the broken lines.

FOOTING DETAIL *(cont.)*

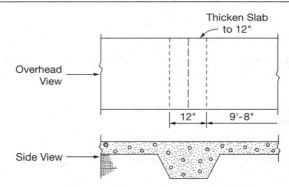

This unusual slab configuration (necessary to support a load-bearing wall) requires detailing so that it is clean to the installers.

FOUNDATION WALL DETAIL

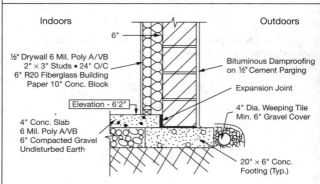

This elevation detail shows all components of a foundation wall.

SLAB DETAIL

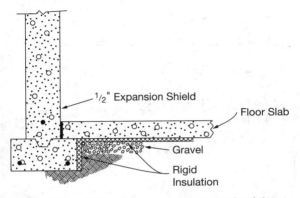

½" Expansion Shield

Floor Slab

Gravel

Rigid Insulation

This elevation detail shows a flexible expansion joint material placed between the concrete slab and the wall.

FLOOR PLAN

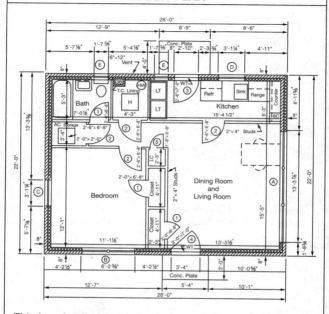

This is a detailed floor plan for a small house. Rooms and structural elements are defined, but plumbing, heating, electrical, and similar systems are omitted.

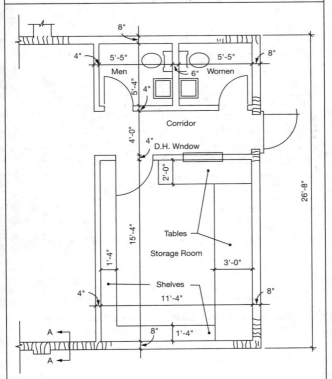

Detailed floor plan of part of a building.
Notice the cutting plane A–A. This defines what the detail A–A
will show. This detail would be an elevation similar to the one
shown on page 4-8.

FLOOR FRAMING PLAN

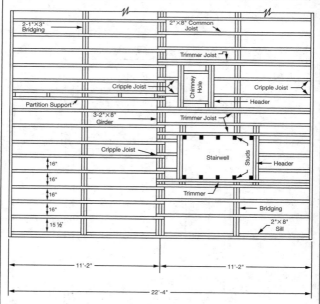

Floor viewed from above with floor surface removed.
Special requirements would be shown in notes.

FLOOR FRAMING DETAIL

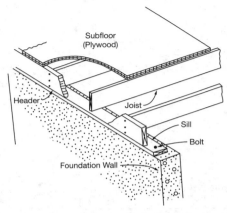

This detail is a perspective drawing, shown not from above or from the side, but from an angle. This view is preferable in some cases but is not used frequently.

FLOORING DETAIL

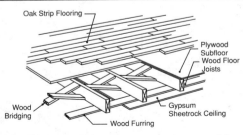

This detail is a perspective drawing of a floor and the ceiling below.

STAIRWAY DETAIL

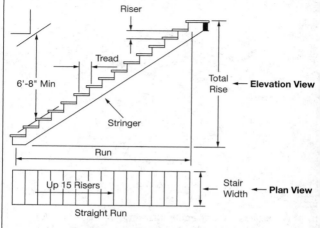

This detail contains both elevation and plan views for full clarity.

FRAMING ELEVATION

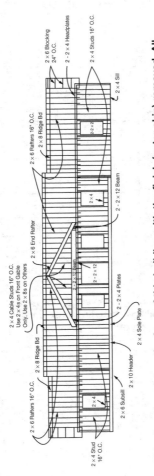

This elevation shows the front of this building with the finish (outer skin) removed. All primary members of the wood frame are shown. Special framing requirements would be shown in notes.

2 × 6 Blocking 24" O.C.

2 - 2 × 4 Headplates

2 × 4 Studs 16" O.C.

2 × 4 Sill

2 × 6 Rafters 16" O.C.

2 × 8 Ridge Bd

2 × 6 End Rafter

2 × 4 Cable Studs 16" O.C. Use 2 × 4s on Front Gable Only. Use 2 × 8s on Others

2 × 8 Ridge Bd

2 × 2

2 - 2 × 12 Beam

2 × 4

2 - 2 × 12 Beam

2 - 2 × 4 Plates

2 - 2 × 12

2 × 6 Rafters 16" O.C.

2 × 4

2 × 4 Sole Plate

2 × 10 Header

2 × 6 Subsill

2 × 4 Stud 16" O.C.

DIFFERING LEVELS

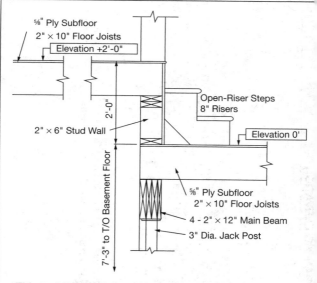

This architectural detail (elevation) shows differing floor levels and exactly how they are to be constructed. Note that T/O means "top of" (basement floor).

DIFFERING LEVELS *(cont.)*

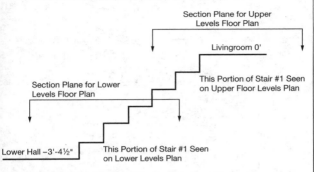

This elevation shows a stairway, along with section plane symbols, showing which areas will be shown on the upper/and lower-level plans.

EXTERIOR ELEVATION DETAIL

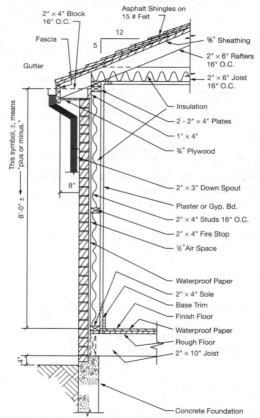

- 2" × 4" Block 16" O.C.
- Asphalt Shingles on 15 # Felt
- Fascia
- 12
- 5
- Gutter
- ⅝" Sheathing
- 2" × 6" Rafters 16" O.C.
- 2" × 6" Joist 16" O.C.
- Insulation
- 2 - 2" × 4" Plates
- 1" × 4"
- ¾" Plywood
- This symbol, ±, means "plus or minus."
- 8'-0" ±
- 8"
- 2" × 3" Down Spout
- Plaster or Gyp. Bd.
- 2" × 4" Studs 16" O.C.
- 2" × 4" Fire Stop
- ½" Air Space
- Waterproof Paper
- 2" × 4" Sole
- Base Trim
- Finish Floor
- Waterproof Paper
- Rough Floor
- 2" × 10" Joist
- 4"
- Concrete Foundation

This sectional view shows construction details that would not be visible in either the plan view or on the elevation view.
Note the depth of detail. For clarity, only the top of the downspout is shown.

4-16

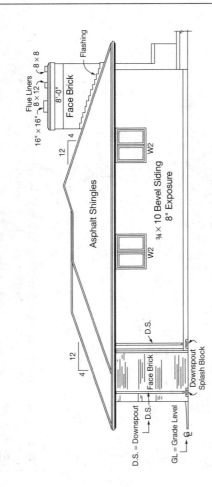

This drawing shows the exterior skin, or finish, of the building. It is an elevation.
Note that all exterior surfaces are specified: siding, shingles, face brick, windows, and flashing.
The roof has a 4/12 pitch, or an angle of about 18 1/2°. Note that window types (W2) are specified.
Reference to a window schedule would tell you the manufacturer or other specific information.

4-17

EXTERIOR ELEVATION

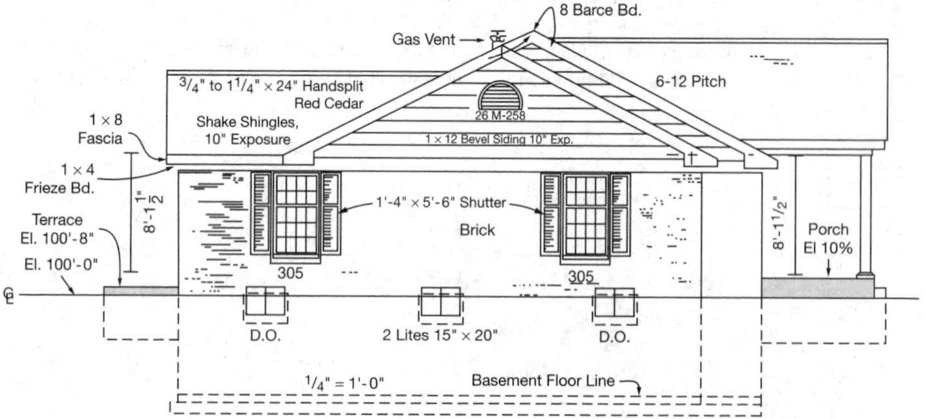

8 Barce Bd.

6-12 Pitch

Gas Vent

Porch EI 10%

8'-11 1/2"

3/4" to 11/4" × 24" Handsplit Red Cedar

Shake Shingles, 10" Exposure

1 × 12 Bevel Siding 10" Exp.

26 M-258

1'-4" × 5'-6" Shutter

Brick

305

D.O.

D.O.

2 Lites 15" × 20"

Basement Floor Line

1/4" = 1'-0"

1 × 8 Fascia

1 × 4 Frieze Bd.

Terrace EI. 100'-8"

EI. 100'-0"

8'-1 1/2"

This is a finish elevation, showing, foundation wall and footing as hidden lines. Notice all the notations, including the term *lites*, which refers to the panels of glass in the ground-level windows.

4-18

SIDING DETAIL

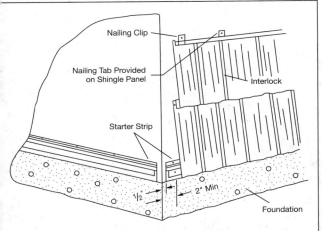

Nailing Clip

Nailing Tab Provided on Shingle Panel

Interlock

Starter Strip

1/2"

2" Min

Foundation

This is a perspective drawing.

SIDING DETAIL *(cont.)*

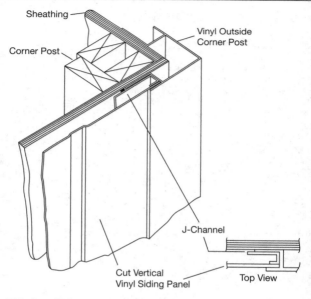

Sheathing

Vinyl Outside
Corner Post

Corner Post

J-Channel

Cut Vertical
Vinyl Siding Panel

Top View

This is a J-channel and cut panel edge installed in a
corner post. The large drawing is a perspective and the
small detail is an overhead view.

CURTAIN WALL DETAIL

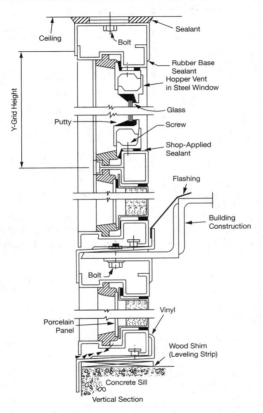

This elevation detail (vertical detail) shows the complexity of an outer window mounting.

Note that most of the window is not shown, so that the detail is not overly large.

ROOF PLAN

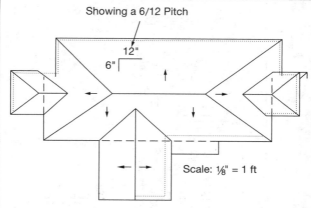

Showing a 6/12 Pitch

12"

6"

Scale: ⅛" = 1 ft

This is a plan showing a pitched, shingle roof for a house.
The arrows show the downward pitch of the roof. The 6/12
pitch notation indicates that the surface rises 6 inches for
every 12 inches of horizontal run, or about 27°.

ROOF PITCH

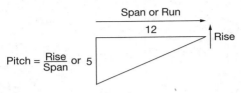

Span or Run

12

$\text{Pitch} = \dfrac{\text{Rise}}{\text{Span}}$ or 5

Rise

The roof pitch indicated above is called a "five-twelve" pitch. A very low-pitched roof would have to be covered with different roofing materials than a steeply pitched roof. Generally, asphalt is used for flatter roofs and shingles for pitched roofs.

FLAT ROOF DETAIL

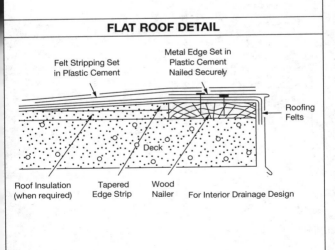

Felt Stripping Set in Plastic Cement

Metal Edge Set in Plastic Cement Nailed Securely

Roofing Felts

Deck

Roof Insulation (when required)

Tapered Edge Strip

Wood Nailer

For Interior Drainage Design

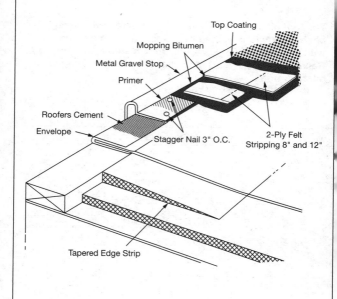

Top Coating

Mopping Bitumen

Metal Gravel Stop

Primer

Roofers Cement

Envelope

Stagger Nail 3" O.C.

2-Ply Felt
Stripping 8" and 12"

Tapered Edge Strip

BUILT-UP ROOF DETAIL

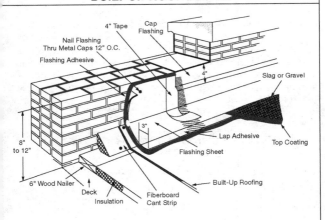

- 4" Tape
- Cap Flashing
- Nail Flashing Thru Metal Caps 12" O.C.
- Flashing Adhesive
- Slag or Gravel
- 4"
- 3"
- Lap Adhesive
- Top Coating
- Flashing Sheet
- Built-Up Roofing
- 8" to 12"
- 6" Wood Nailer
- Deck
- Insulation
- Fiberboard Cant Strip

FLASHING DETAIL

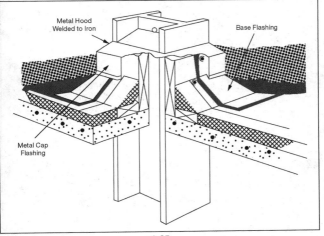

- Metal Hood Welded to Iron
- Base Flashing
- Metal Cap Flashing

GRAPHIC SYMBOLS

The symbols shown on the following pages seem to be the most common and acceptable, judged by their frequency of use in architectural offices. This list can be and often is expanded to include other symbols generally used, even if not shown here. Adoption of these symbols as standard practice is desirable to improve communication in the industry.

DRAWING CONVENTIONS AND SYMBOLS

$\underset{11^{1}/_{2}\ T}{\text{Up 17R}}$ → Stair direction symbol

North point;
to be placed on each
floor plan, generally in
lower right-hand corner
of drawings

Indication arrows
drawn with straight
lines (not curved);
must touch object

DRAWING CONVENTIONS AND SYMBOLS *(cont.)*

Indicates section number

Indicates drawing sheet
on which section is shown

C
A-3

C
A-3

3
A-1

7
A-5

II
A-3

Section Lines and Section References

Indicates detail number

5
A-8

9
A-4

II

Indicates drawing sheet
on which detail is shown

Detail References

DRAWING CONVENTIONS AND SYMBOLS (cont.)

Dash and dot

Center lines, projections, existing elevations lines

Dash and double dot line

Property lines, boundary lines

Dotted line

Hidden, future or existing construction to be removed

Break line

To break off parts of drawing

Linework

4' 0"	8"	Slash
2' 8"	4"	
8' 1/2"	6³/₄"	Arrow
26' 8"	2"	Dot
5' 4"	1/2"	Accent

Horizontal   4' 0"   6' 2"   Vertical

Dimension Lines

461.0'	New or required point elevation
+ 461.0'	Existing point elevation (plan)
268	Existing contours elevation noted on high side
320	New contours elevation noted on high side
TB-1	Test boring
C / A-9	Building section Reference drawing number
7 / A-11	Wall section or elevation Reference drawing number
7 / A-12	Detail Reference drawing number
1302	Room/space number
354	Equipment number

	Match line shaded portions— the side considered
	Level line control point or datum
3	Revision
E	Window type
123 B	Door number (if more than one door per room subscript letters are used)
A — 4	Column reference grids
N / Mag north	Project north (magnetic north arrow used on plot site plan only)

ARCHITECTURAL SYMBOLS

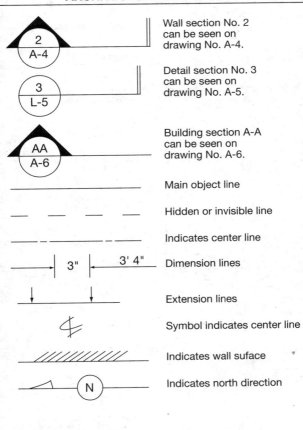

Wall section No. 2 can be seen on drawing No. A-4.

Detail section No. 3 can be seen on drawing No. A-5.

Building section A-A can be seen on drawing No. A-6.

Main object line

Hidden or invisible line

Indicates center line

Dimension lines

Extension lines

Symbol indicates center line

Indicates wall suface

Indicates north direction

ARCHITECTURAL SYMBOLS (cont.)

(ca) — – – –	Column line grid
⟨5⟩— or ▷5—	Partition type
⟨A⟩	Window type
(05)	Door number
⌐05⌐	Room number
(10'-0")	Ceiling height
△2 ⌒⌒	Revision marker
——→‖←—	Break in a continuous line
(3) —→	Refer to note #3
◐ 100'-0" – – –	Elevation marker
◇ with 1/4 A-5 2/3	Interior elevations 1, 2, 3, & 4 can be seen on drawing A-5. Direction of triangle indicates elevation.

SYMBOLS FOR STAIRS

Stairs Going Up

Stairs Going Down

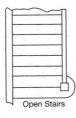

Open Stairs

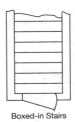

Boxed-in Stairs

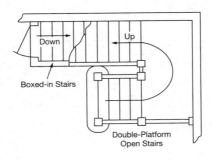

Boxed-in Stairs

Double-Platform
Open Stairs

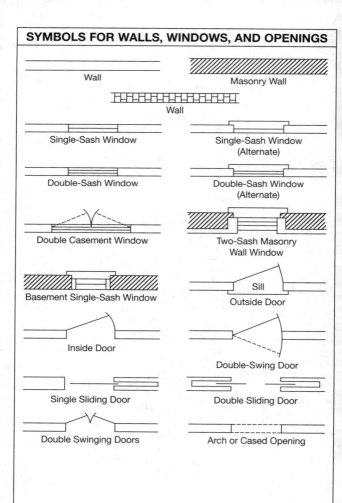

SYMBOLS FOR WALLS, WINDOWS, AND OPENINGS

Wall

Masonry Wall

Wall

Single-Sash Window

Single-Sash Window
(Alternate)

Double-Sash Window

Double-Sash Window
(Alternate)

Double Casement Window

Two-Sash Masonry
Wall Window

Basement Single-Sash Window

Sill

Outside Door

Inside Door

Double-Swing Door

Single Sliding Door

Double Sliding Door

Double Swinging Doors

Arch or Cased Opening

DOOR AND WINDOW SYMBOLS

Door Symbols

Type	Symbol

Single-Swing with Threshold in Exterior Masonry Wall

Single Door, Opening In

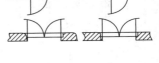

Double Door, Opening Out

Single-Swing with Threshold in Exterior Frame Wall

Single Door, Opening Out

Double Door, Opening In

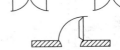

Refrigerator Door

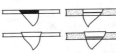

Window Symbols

Type | Symbol

| | Wood or Metal Sash in Frame Wall | Metal Sash in Masonry Wall | Wood Sash in Masonry Wall |

Double Hung

Casement

Double, Opening Out

Single, Opening In

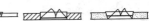

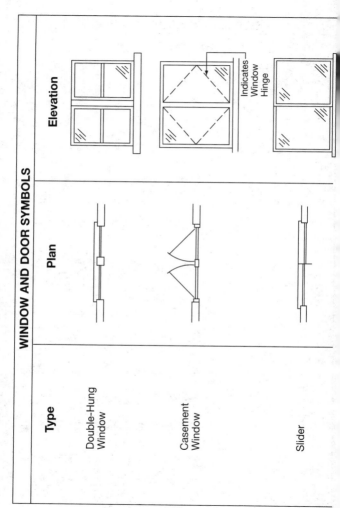

WINDOW AND DOOR SYMBOLS

Type	Plan	Elevation
Double-Hung Window		
Casement Window		Indicates Window Hinge
Slider		

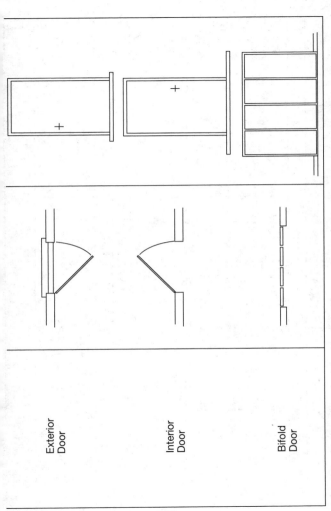

Exterior
Door

Interior
Door

Bifold
Door

DOOR SCHEDULE

Symbol	Quantity	Type	Door Size	Remarks
A	1	Panel	3'- 0" × 6'- 8" × 1¾"	Fir
B	1	Flush	2'- 6" × 6'- 8" × 1¾"	Birch, solid core
C	1	Panel	2'- 6" × 6'- 8" × 1¾"	Fir w/2 hammered glass lites
D	1	Panel	2'- 6" × 6'- 8" × 1¾"	Fir w/ventilating lite
E	1	Panel	2'- 6" × 6'- 8" × 1¾"	Fir
F	2	French	2'- 6" × 6'- 8" × 1¾"	Fir
G	1	Glass Sliding	9'- 0" × 6'- 10"	Aluminum
H	2	Flush	2'- 0" × 6'- 8" × 1¾"	Masonite, w/louvres
J	4	Louvre	1'- 4" × 6'- 8" × 1⅜"	Fir
K	3	Bifold Louvre	5'- 6" × 6'- 8" × 1⅜"	Fir, track at top only
L	4	Panel	2'- 6" × 6'- 8" × 1⅜"	Fir
M	2	Panel	1'-10" × 6'- 8" × 1⅜"	Fir
N	1	Flush	2'- 6" × 6'- 8" × 1⅜"	Birch, hollow core w/louvre
O	2	Panel	2'- 4" × 6'- 8" × 1⅜"	Fir
P	1	Pocket Louvre	2'- 0" × 6'- 8" × 1⅜"	Fir
R	1	Panel	2'- 0" × 6'- 8" × 1⅜"	Fir
S	1	Panel Dutch	2'- 6" × 6'- 8" × 1⅜"	Fir

WINDOW SCHEDULE

Symbol	Quantity	Type	Window Size	Remarks
1	2	Sliding	4'- 0" × 4'- 0"	Aluminum
2	1	Sliding	4'- 0" × 3'- 0"	Aluminum
3	1	Sliding	2'- 0" × 3'- 0"	Aluminum
4	4	Ventilating	6'- 0" × 6'- 10"	Aluminum
5	1	Fixed	6'- 0" × 6'- 10"	Aluminum
6	1	Ventilating	2'- 8" × 6'- 10"	Aluminum

SYMBOLS FOR CHIMNEYS AND FIREPLACES

Plain Chimney

Chimney with Tile

Plain Chimney Double Flue

Fireplace with Flue from Below

Flue

Fireplace with Ash Dump

Ash
Dump

Corner Fireplace with Tile

Tile

MATERIALS SYMBOLS

Material	Plan	Elevation	Section
Wood	Floor Areas Left Blank	Siding / Panel	Finish / Framing
Brick	Face / Common	Face or Common	Same as Plan View
Stone	Cut / Rubble	Cut / Rubble	Cut / Rubble

4-40

Concrete			Same as Plan View
Concrete Block			Same as Plan View
Earth	None	None	
Glass			Large Scale Small Scale
Insulation	Same as Section	Insulation	Loose Fill or Batt Board

4-41

MATERIALS SYMBOLS (cont.)

Material	Plan	Elevation	Section
Plaster	Same as Section	Plaster	Stud / Lath and Plaster
Structural Steel	— — —	Indicate by Note	
Sheet Metal Flashing	Indicate by Note		Show Contour

4-42

Material	Symbol	Floor	Wall
Tile			
Porous Fill		None	None
Plywood		Indicated by Note	Indicated by Note

4-43

MATERIALS SYMBOLS *(cont.)*

Material	Plan	Elevation	Section
Batt Insulation		None	Same as Plan
Rigid Insulation		None	Same as Plan
Glass			Small Scale Large Scale

Gypsum Wallboard			Same as Plan
Acoustical		None	
Ceramic Wall Tile			Same as Plan
Floor Tile		None	

4-45

MATERIALS SYMBOLS (cont.)

Glass

Glass	Structural	Glass Block

Insulation

Batt/Loose Fill	Rigid	Spray/Foam

Finishes

Acoustical Tile	Ceramic Tile – Large Scale	Ceramic Tile – Small Scale
Carpet and Pad	Gypsum Wallboard	Metal Lath and Plaster

Finishes (cont.)

Plastic	Resilient Flooring/Plastic Laminate	Terrazzo

Plan and Section Indications
Partition Indications

Wood Stud	Metal Stud	Special Finish Face

Elevation Indications

Brick	Ceramic Tile	Concrete/Plaster
Glass	Sheet Metal	Shingles/Siding

4-47

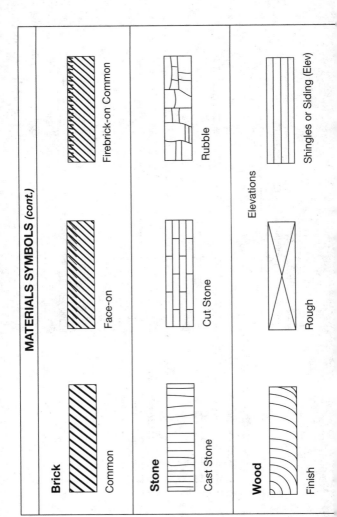

MATERIALS SYMBOLS (cont.)

Brick

Common — Face-on — Firebrick-on Common

Stone

Cast Stone — Cut Stone — Rubble

Wood

Finish — Rough — Shingles or Siding (Elev)

Elevations

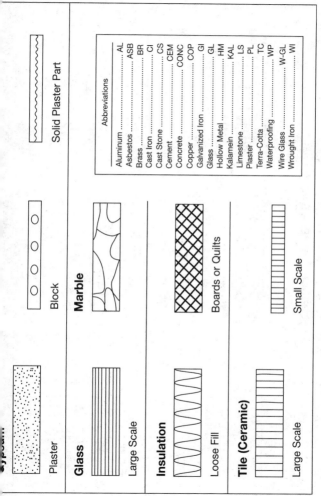

Plaster

Block

Solid Plaster Part

Glass

Large Scale

Marble

Insulation

Loose Fill

Boards or Quilts

Tile (Ceramic)

Large Scale

Small Scale

Abbreviations	
Aluminum	AL
Asbestos	ASB
Brass	BR
Cast Iron	CI
Cast Stone	CS
Cement	CEM
Concrete	CONC
Copper	COP
Galvanized Iron	GI
Glass	GL
Hollow Metal	HM
Kalamein	KAL
Limestone	LS
Plaster	PL
Terra-Cotta	TC
Waterproofing	WP
Wire Glass	W-GL
Wrought Iron	WI

4-49

MATERIALS SYMBOLS (cont.)

Concrete

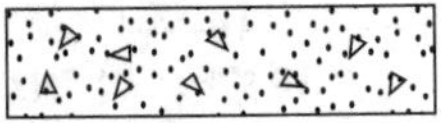

Stone

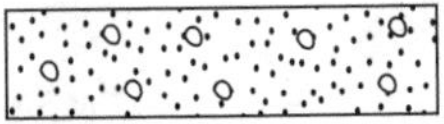

Cement

Cinder

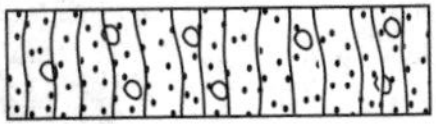

Terrazzo

Cast Block

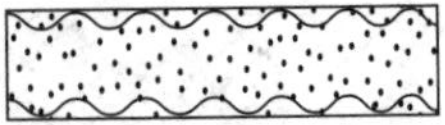

Earth, etc.

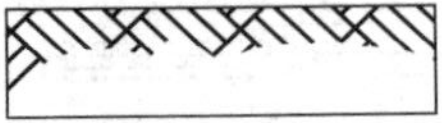

Earth

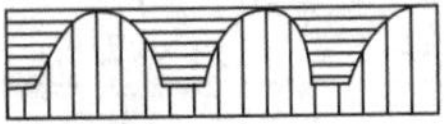

Rock

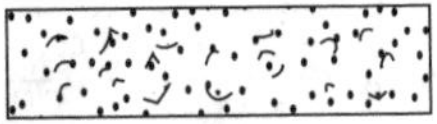

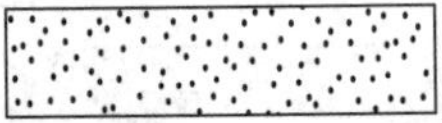

Sand

Cinders

4-50

Metals

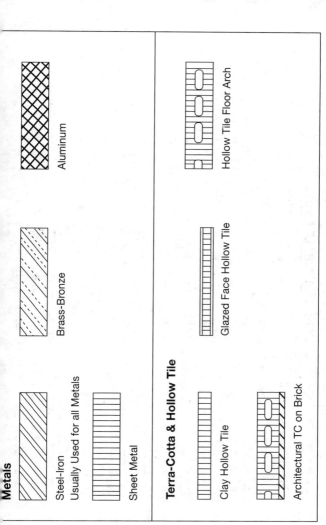

Steel-Iron
Usually Used for all Metals

Sheet Metal

Brass-Bronze

Aluminum

Terra-Cotta & Hollow Tile

Clay Hollow Tile

Glazed Face Hollow Tile

Architectural TC on Brick

Hollow Tile Floor Arch

MATERIALS SYMBOLS *(cont.)*

Plans of Exterior Walls

Exterior Side of Wall

Brick

Face Brick

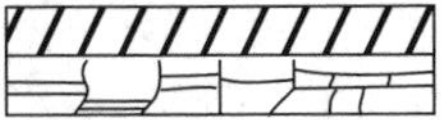

Rubble

Interior

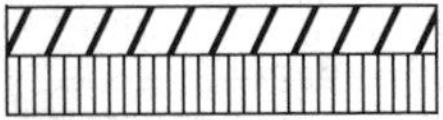

Clay Hollow Tile

Cut Stone

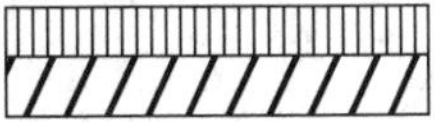

Architectural TC

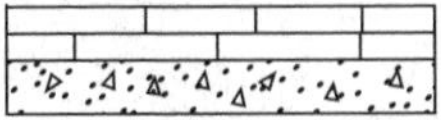

Concrete-Stone

Brick

Exterior Side of Wall

Cut Stone

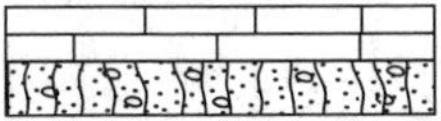

Rubble

Concrete Block

Glazed Face Hollow Tile

Interior

Cast Stone

Brick

Clay Hollow Tile

Gypsum Block

Concrete Block

Brick-Plastered

Solid Plaster

Glazed Face Hollow Tile

Stud

4-53

MATERIALS SYMBOLS (cont.)

Sections of Floor Finishes

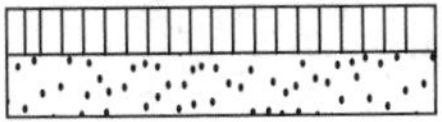

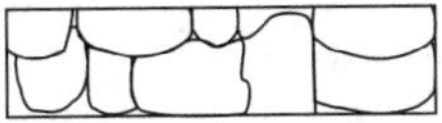

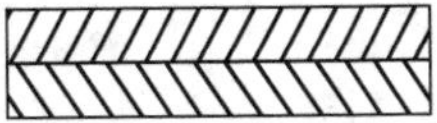

Wood

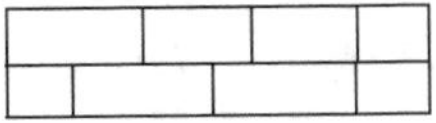

Stone

Marble

Cement

Tile

Terrazzo

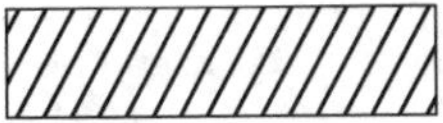

Brick

CHAPTER 5
Structural Drawings

STRUCTURAL DRAWINGS

Structural pages are prepared by engineers (not architects) and show elements such as structural steel and reinforced concrete.

Structural pages may not be prepared for simple projects that have no structural steel, reinforced concrete, or other structural systems. For example, a set of blueprints for a simple wood-framed house does not require structural pages.

STRUCTURAL MEMBERS

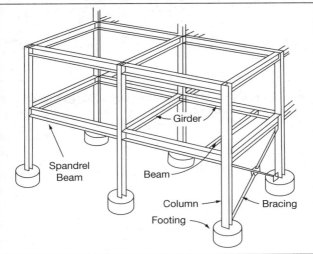

Spandrel Beam

Girder

Beam

Column

Bracing

Footing

OPEN-WEB JOIST PLAN

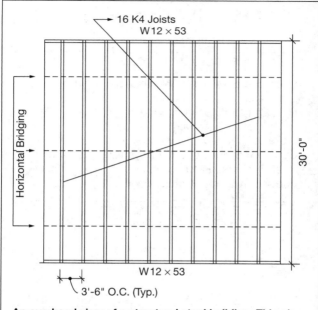

16 K4 Joists
W 12 × 53

Horizontal Bridging

30'-0"

W 12 × 53

3'-6" O.C. (Typ.)

An overhead view of a structural steel building. This plan shows what will be supporting the floor.

5-2

ROOF FRAMING PLAN

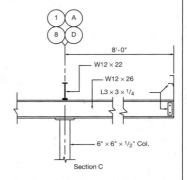

A roof framing plan for structural steel construction.
The inset shows Section C in detail.

5-3

COLUMN DETAILS

Top Views   **Top Views**

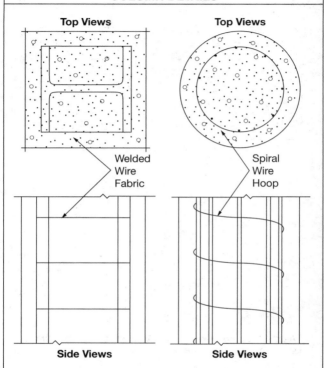

Welded
Wire
Fabric

Spiral
Wire
Hoop

Side Views   **Side Views**

These column details give both overhead and side views, for clarity.

COLUMN SCHEDULE

Column	Size	Base Plate	Setting Plate	Cap. Plate
A1				$5" \times \frac{5}{8}" \times 10"$
A2				$5" \times \frac{5}{8}" \times 10"$
A3				$5" \times \frac{5}{8}" \times 10"$
B1				Thru Plate
B2	Typical 4" ∅ Std. Col. × 10.79 #/Ft.	Typical 9" × $\frac{3}{4}$" × 9" Base Plate	Typical 9" × $\frac{1}{4}$" × 9" Setting Plate	$5\frac{1}{2}" \times \frac{5}{8}" \times 12"$
B3				$5\frac{1}{2}" \times \frac{5}{8}" \times 12"$
C1				Thru Plate
C2				$5\frac{1}{2}" \times \frac{5}{8}" \times 12"$
C3				$5\frac{1}{2}" \times \frac{5}{8}" \times 12"$
D1				Thru Plate
D2				$5\frac{1}{2}" \times \frac{5}{8}" \times 12"$
D3				$5\frac{1}{2}" \times \frac{5}{8}" \times 12"$
E1				$5" \times \frac{5}{8}" \times 10"$
E2				$5\frac{1}{2}" \times \frac{5}{8}" \times 12"$
E3				$5\frac{1}{2}" \times \frac{5}{8}" \times 12"$

COMPLEX COLUMN DETAIL

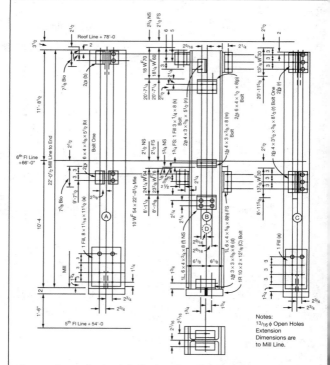

A complex column detail, showing the notation and arrangement normally employed on drawings of this type.

REBAR MARKINGS

Line System—Grade Marks

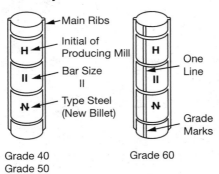

Grade 40
Grade 50

Grade 60

Number System—Grade Marks

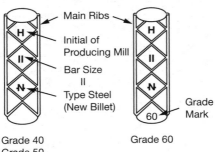

Grade 40
Grade 50

Grade 60

N = New Billet   **I** = Rail

A = Axle   **W** = Low Alloy

STRUCTURAL STEEL SYMBOLS

Descriptive Name	Shape	Identifying Symbol	Typical Designation Height wt/ft in lb	Nominal Size Height Width
Wide Flange Shapes	I	W	W21 × 142	21 × 13
Miscellaneous Shapes	I	M	M8 × 6.5	8 × 2¼
American Standard Beams	I	S	S8 × 23	8 × 4
American Standard Channels	[	C	C6 × 13	6 × 2
Miscellaneous Channels	[	MC	MC8 × 20	8 × 3
Angles—Equal Legs	L	L	L6 × 6 × ½*	6 × 6
Angles—Unequal Legs	L	L	L8 × 6 × ½*	8 × 6
Bulb Angles	L	BL	BL6 × 3½ × 17.4	3½ × 6
Structural Tees (cut from wide flange)	T	WT	WT12 × 60	12
Structural Tees (cut from miscellaneous shapes)	T	MT	MT5 × 4.5	5
Structural Tees (cut from am. std. beams)	T	ST	ST9 × 35	9
Tees	T	T	T5 × 11.5	3 × 5
Wall Tee	T	AT	AT8 × 29.2	4⅞ × 7¾
Elevator Tees	T	ET	ET4 × 24.5	4⅛ × 5½
Zees	⌐	Z	Z4 × 15.9	6 × 3½

*Size only

CHAPTER 6
Plumbing Drawings

PLUMBING DRAWINGS

Plumbing drawings are sometimes included among the mechanical pages in a set of blueprints, but are just as often provided separately and numbered P1, P2, P3, and so on.

Plumbing drawings include floor plans showing the locations of sinks, toilets, and so on, details, and a specialized type of drawing called an *isometric*. An isometric is a type of perspective drawing showing the approximate routing of the plumbing pipes in three dimensions.

Plumbing work may also be shown on site plans. These plans normally include plumbing between buildings or service to the building.

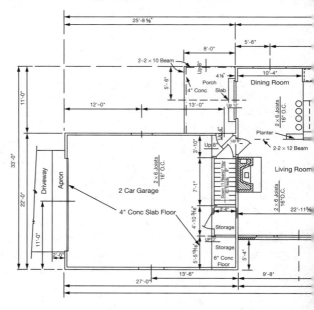

A floor plan showing plumbing fixtures for a house.

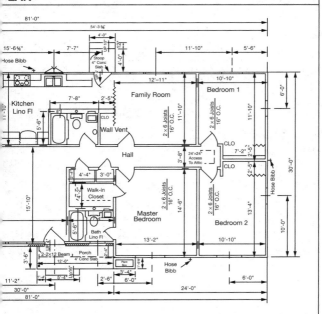

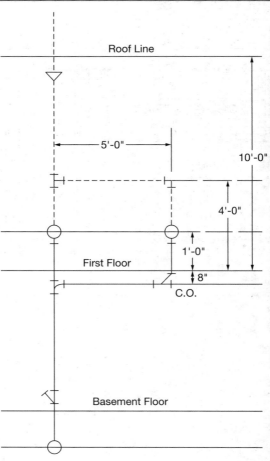

Roof Line

5'-0"

10'-0"

4'-0"

First Floor

1'-0"

8"

C.O.

Basement Floor

6-4

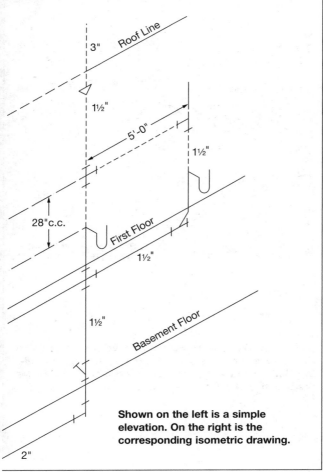

3"

Roof Line

1½"

5'-0"

1½"

28"c.c.

First Floor

1½"

1½"

Basement Floor

2"

Shown on the left is a simple elevation. On the right is the corresponding isometric drawing.

ISOMETRIC

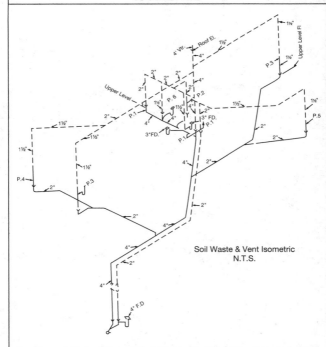

Soil Waste & Vent Isometric
N.T.S.

An isometric is a specialized perspective drawing showing the approximate paths of plumbing pipes through a building. Notice that pipes on the upper level and the roof level are specified. N.T.S. signifies "Not to scale."

PLUMBING FIXTURE SCHEDULE

Mark	Fixture	Pipe Size					Remarks
		Trap	Waste	Vent	HW	CW	
P-1	Water Closet	Int.	4"	2"	—	½"	Floor Mounted Tank Type Handicapped
P-2	Lavatory	1½"	1½"	1½"	½"	½"	Wall Hung Handicapped
P-3	Hand Sink	1½"	1½"	1½"	½"	½"	Furnished by Others
P-4	Sink/Washer						
P-5	Rinse Sink	↓	↓	↓	↓	↓	
P-6	Pot Sink (3 Compart.)	3'@2"	3'@2"				
P-7	Janitors Mop Sink	3"	3"	2"	½"	½"	
P-8	Bar Sink	1½"	1½"	1½"	½"	½"	↓ ↓ ↓

Plumbing fixture schedules are used to specify details that don't fit on plan sheets. Notice that many of the plumbing fixtures are supplied by others.

PLUMBING FIXTURE DETAIL

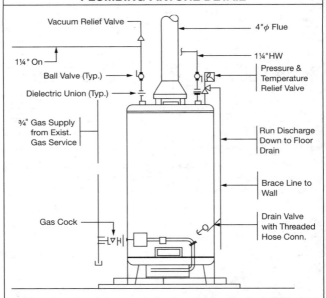

Vacuum Relief Valve

4"⌀ Flue

1¼" On

1¼" HW

Ball Valve (Typ.)

Pressure & Temperature Relief Valve

Dielectric Union (Typ.)

¾" Gas Supply from Exist. Gas Service

Run Discharge Down to Floor Drain

Brace Line to Wall

Gas Cock

Drain Valve with Threaded Hose Conn.

Elevation detail for a gas-fired water heater installation.

STANDARD FLOOR PLAN

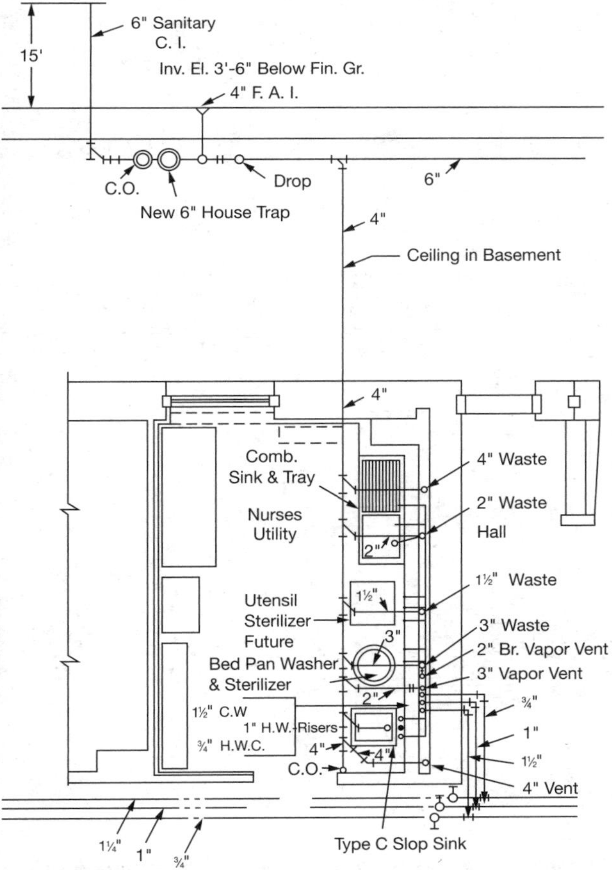

6" Sanitary C. I.

Inv. El. 3'-6" Below Fin. Gr.

15'

4" F. A. I.

C.O.

New 6" House Trap

Drop

6"

4"

Ceiling in Basement

4"

Comb. Sink & Tray

Nurses Utility

4" Waste

2" Waste

Hall

1½" Waste

Utensil Sterilizer Future

Bed Pan Washer & Sterilizer

3" Waste

2" Br. Vapor Vent

3" Vapor Vent

1½" C.W

1" H.W. Risers

¾" H.W.C.

C.O.

¾"

1"

1½"

4" Vent

Type C Slop Sink

1¼" 1" ¾"

The bottom drawing is a standard floor plan, and the top drawing is an elevation. This is an unusual type of floor plan.

6-9

SHOP DRAWINGS

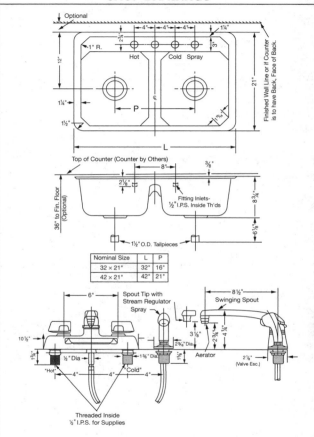

These shop drawings are very precise, detailed drawings, typically provided by manufacturers.

COMMON PLUMBING MEASUREMENT TERMS

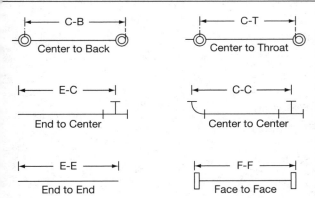

Measurement terms that often appear on plumbing plans, especially on details.

PIPING SYMBOLS

Soil, Waste, or Leader (above grade)	———————
Sanitary above Grade	——— S ———
Soil, Waste, or Leader (below Grade)	— — — —
Sanitary below Grade	– – – – S – – – –
Vent	— — — —
Combination Waste and Vent	——— CWV———
Waste Line	——— W ———
Drain Line	——— D ———
Vent Line	——— V ———
Indirect Waste	——— IW ———
Indirect Vent	——— IV ———

Sanitary Sewer	——— SS ———
Storm Drain	——— SD ———
Storm above Grade	——— ST ———
Storm below Grade	– – – ST – – –
Rain Water Leader	——— RWL ———
Sump Pump Discharge	——— SPD ———
Service Water	——— SWS ———
Cold Water	——— · ——— · ———
Soft Cold Water	——— SWC ———
Drinking Water Supply	——— DWS ———
Drinking Water Return	——— DWR ———
Chilled Drinking Water Supply	——— CDWS ———

PIPING SYMBOLS (cont.)

Chilled Drinking Water Return	—— CDWR ——	Tempered Water Supply	—— TWS ——
Hot Water	—— ·· —— ·· ——	Tempered Water Return	—— TWR ——
Hot Water Circulation	___ _ _ ___	Acid Water	—— ACID ——
Hot Water Return	— ·· — ·· —	Acid Waste above Grade	—— AW ——
Sanitizing Hot Water Supply (180°F)	—+ ·· —+ ·· —+	Acid Waste below Grade	_ _ _AW _ _ _
Sanitizing Hot Water Return (180°F)	—+ ··· —+ ··· —+	Acid Vent	_ _ _ AV _ _ _
Industrialized Hot Water Supply	___ IHW ___	Chemical-Resistant Waste	—— CRW——
Industrialized Hot Water Return	—— IHR ——	Fire Line	— F —— F —
Industrialized Cold Water	—— ICW——	Wet Standpipe	—— WSP——
Industrial Waste	—— INW ——	Dry Standpipe	—— DSP——
Tempered Water, Potable	—— T ——	Combination Standpipe	___ CSP___
		Main Sprinkler Supply	—— SPR——

6-13

PIPING SYMBOLS *(cont.)*

Branch and Head Sprinkler	—o———o—	Hydrogen	— H —
Gas—Low Pressure	— G — G —	Helium	— HE —
Gas—Medium Pressure	— MG —	Argon	— AR —
Gas—High Pressure	— HG —	Liquid Petroleum Gas	— LPG —
Compressed Air	— CA —	Pneumatic Tubes Tube Runs	— PN —
Vacuum	— VAC —	Cast Iron	— CI —
Vacuum Cleaning	— VC —	Culvert Pipe	— CP —
Oxygen	— O —	Clay Tile	— CT —
Liquid Oxygen	— LO —	Ductile Iron	— DI —
Nitrogen	— N —	Reinforced Concrete	— RCP —
Liquid Nitrogen	— LN —	Drain—Open Tile or Agricultural Tile	= = = =
Nitrous Oxide	— NO —	Low-Pressure Steam	— LPS —

PIPING SYMBOLS (cont.)

Medium-Pressure Steam	—— MPS ——		Medium-Temperature Hot Water Return	—— MTWR ——
High-Pressure Steam	—— HPS ——		High-Temperature Hot Water Return	—— HTWR ——
Low-Pressure Return	—— LPR ——		Boiler Blowdown	—— BBD ——
Medium-Pressure Return	—— MPR ——		Feedwater Pump Discharge	—— FPD ——
High-Pressure Return	—— HPR ——		Hot Water Heating Supply	—— HWS ——
Low-Temperature Hot Water Supply	—— HWS ——		Hot Water Heating Return	—— HWR ——
Medium-Temperature Hot Water Supply	—— MTWS ——		Fuel Oil Suction	—— FOS ——
High-Temperature Hot Water Supply	—— HTWS ——		Fuel Oil Return	—— FOR ——
			Fuel Oil Tank Vent	—— FOV ——
Low-Temperature Hot Water Return	—— HWR ——		Existing Piping	—— (NAME) E ——
			Pipe to Be Removed	✕✕ (NAME) ✕✕

VALVE SYMBOLS		
Type	**Screwed**	**Bell and Spigot**
Gate Valve		
Globe Valve		
Check Valve		
Angle Check Valve		
Stop Cock		
Relief or Safety Valve		

VALVE SYMBOLS *(cont.)*

Angle Globe Valve		Solenoid Valve	
Angle Gate Valve		Diaphragm-Operated Valve	
Quick-Opening Valve		Reducing Valve (Self-Actuated)	
Float-Opening Valve		Reducing Valve (External Pilot Connection)	
Post Indicator Gate Valve		Lock Shield	
Plug Valve		Two-Way Automatic Control	
Plug Cock		Three-Way Automatic Control	
Butterfly Valve		Gas Cock	
Pressure-Reducing Valve		Shock Absorber	
Hose Gate		OS & Y Gate	
Three-Way Valve		Strainer	
Motor-Operated Valve		Temperature- and Pressure-Relief Valve	
Motor-Operated Gate Valve			

FITTING SYMBOLS

Fitting connections are often specified with a fitting symbol. For example, an elbow could have the following types of connections:

Flanged		Soldered	
Welded		Bell and Spigot	
Screwed		Solvent Cement	

Fittings are shown with screwed connections unless specified.

Bushing		Reducer, Concentric	
Cap		Reducer, Concentric, Straight Invert	
Connection, Bottom		Reducer, Concentric, Straight Crown	
Connection, Top			
Elbow, Reducing, Showing Sizes		Tee, Reducing, Showing Sizes	
		Tee, Single Sweep	
Elbow, Double Branch		Union	

FITTING SYMBOLS *(cont.)*		
Type	**Screwed**	**Bell and Spigot**
Joint, Coupling		
Elbow—90°		
Elbow—45°		
Elbow—Turned Up		
Elbow—Turned Down		
Elbow—Long Radius		

FITTING SYMBOLS *(cont.)*		
Type	**Screwed**	**Bell and Spigot**
Side Outlet Elbow—Outlet Down		
Side Outlet Elbow—Outlet Up		
Base Elbow		
Double-Branch Elbow		
Single-Sweep Tee		
Double-Sweep Tee		

FITTING SYMBOLS *(cont.)*

Type	Screwed	Bell and Spigot
Reducing Elbow, Showing Sizes		
Tee		
Tee—Outlet Up		
Tee—Outlet Down		
Side-Outlet Tee—Outlet Up		
Side-Outlet Tee—Outlet Down		

FITTING SYMBOLS *(cont.)*

Type	Screwed	Bell and Spigot
Cross		
Reducer		
Eccentric Reducer		
Lateral		
Expansion Joint Flanged		

SPECIALTY AND MISCELLANEOUS SYMBOLS

Alignment Guide, Pipe		Flanged Joint	
Anchor, Intermediate	PA	Flexible Connection	
Anchor, Main	PA	Flexible Connector	
Ball Joint		Flow Direction	
Concentric Reducer		Flowmeter, Orifice	OFM-1
Eccentric Reducer		Flowmeter, Venturi	VFM-1
Elbow Looking Up		Flow Switch	FS
Elbow Looking Down		Funnel Drain, Open	
Expansion Joint		Hanger, Rod	H
Expansion Loop		Hanger, Spring	H

6-23

SPECIALTY AND MISCELLANEOUS SYMBOLS (cont.)

Pitch of Pipe, Rise (R) Drop (D)	Thermometer
Pressure Gauge	Thermostat
Pressure Switch	Thermostat, Self-Contained
Pressure Switch, Dual (high–low)	Thermostat, Remote Bulb
Pump, Indicating Use	Thermostatic Trap
Spool Piece, Flanged	Float and Thermostatic Trap
Strainer	House Trap
Strainer, Blow-Off	P-trap
Strainer, Duplex	Traps, Steam, Indicating Type
Tank, Indicating Use	'Y'

FIXTURE SYMBOLS

Baths

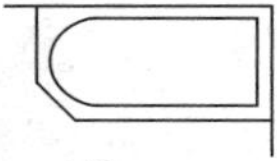

Corner

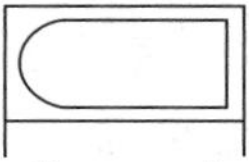

Recessed

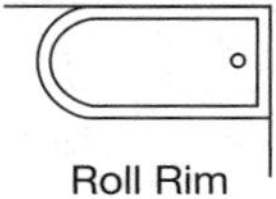

Roll Rim

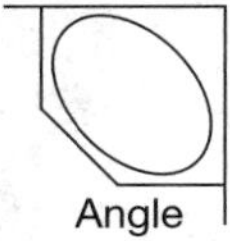

Angle

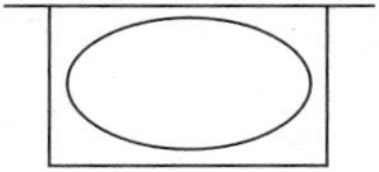

Whirlpool

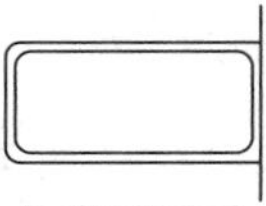

Institutional or Island

SB

Sitz Bath

FB

Foot Bath

Showers

Stall

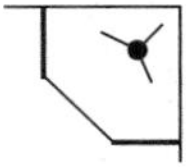

Corner Stall

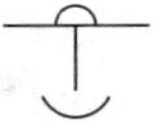

(Plan) 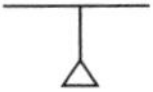   (Elev.)

Shower Head

(Plan)

(Elev)

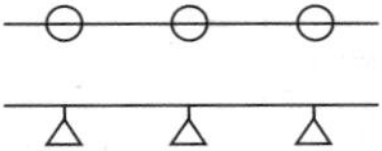

Overhead Gang
Shower Heads

FIXTURE SYMBOLS (cont.)

Water Closets

Floor	Hung	Bidet	Low Tank	No Tank

Urinals

Wall	Stall	Corner	Trough	Pedestal

Lavatories

Vanity	Wall	Pedestal
Corner	Manicure/Medical	Dental

Drinking Fountains/Electric Water Coolers

Floor or Wall	Recessed	Semirecessed	Pedestal
Wall	Wall Mounted	Trough	

FIXTURE SYMBOLS (cont.)

Sinks/Dishwashers

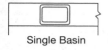

Single Basin

Right Drainboard

Twin Basin

Left Drainboard

Plain Kitchen Sink

Double Drainboard

Sink/Dishwasher
Combination

DW

Dishwasher

Wash Sink

Wash Sink Wall Type

Service Sinks

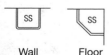

SS	SS
Wall	Floor

Wash Fountains

WF	WF
Circular	Semicircular

FIXTURE SYMBOLS *(cont.)*

Laundry Trays

Single

Double

Combination Sink and
Laundry Tray

Hot Water

Heater

Tank

Separators

Gas

Oil

Cleanouts

Cleanout

Cleanout

Floor

Wall

Drains

Drain

Floor Drain

Floor Drain
with Backup Valve

Garage Drain

Roof Drain

Roof Sump

Miscellaneous

Wall Casting

Wall Hydrant
or Siamese

Manhole
(identify by number)

Inlet Basin
(identify by number)

Catch Basin
(identify by number)

Meter

Hose Rack

HR

Hose Bib

HB

Gas Outlet

G

Vacuum Outlet

FIRE PROTECTION SYMBOLS

Piping

Fire Protection Water Supply	—— F ——	Pendant Heads	●————●
Wet Standpipe	—— WSP ——	Flush-Mounted Heads	⊗————⊗
Dry Standpipe	—— DSP ——		
		Sidewall Heads	▼————▼
Combination Standpipe	—— CSP ——		
		Fire Hydrant	
Automatic Fire Sprinkler	—— SP ——		
Drain	—— D ——	Wall Fire Department Connection	
Riser and Branch (show size)	⊗—o—o 4 in.	Sidewalk Fire Department Connection	
Pipe Hanger	—/—		
Control Valve	—T—	Fire Hose Rack	FHR
Alarm Check Valve		Surface-Mounted Fire Hose Cabinet	FHC
Dry Pipe Valve	◆		
Upright Fire Sprinkler Heads	o————o	Recessed Fire Hose Cabinet	FHC

FIRE PROTECTION SYMBOLS (cont.)

Signal Detectors

Heat (thermal)	⊙	Flame	⊘
Smoke	⊙	Control Panel	☐—FCP
Gas	▲		

Valves

Air Line	——⊘——	Globe	——▶◀—
Ball	——│○│——	Globe, Angle	▶
Butterfly	——│││——	Globe, Stop Check	——▶◀—
Diaphragm	——◁▷——	Plug Valve	——▽——
Gate	——◁▷——	Three-Way	——◁▷—
Gate, Angle	△		

Valve Actuators

Nonrising Steam	⊤	Gear	G
Outside Stem and Yoke	+	Motor	M
Lever	⌐	Solenoid	S

FIRE PROTECTION SYMBOLS (cont.)

Special-Duty Valves

Check, Swing Gate

Check, Spring

Control, Electric-Pneumatic

Control, Pneumatic-Electric

Hose End Drain

Lock Shield

Needle

Pressure Reducing, Self-Contained

Pressure Reducing, External Pressure

Pressure Reducing, Differential Pressure

Quick Opening

Quick Closing, Fusible Link

Relief (R) or Safety (S)

Solenoid

Square Head Cock

CHAPTER 7
Mechanical Drawings

MECHANICAL DRAWINGS

The systems covered by mechanical drawings include plumbing, piping, heating, ventilation, air conditioning, refrigeration, and similar mechanical systems.

Mechanical pages are prepared by certified professional mechanical engineers, and each page should be sealed by the engineer who prepared them. Mechanical pages usually account for a significant portion of the blueprint set.

Electrical control systems and related devices may be shown on the mechanical sheets of a set of blueprints. Electrical motors, disconnect switches, and motor starters may also be shown on these sheets.

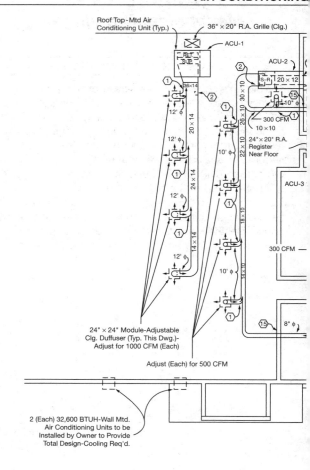

Roof Top-Mtd Air Conditioning Unit (Typ.)

36" × 20" R.A. Grille (Clg.)

ACU-1

ACU-2

S-R 20 × 12

300 CFM
10 × 10

24"× 20" R.A.
Register
Near Floor

ACU-3

300 CFM

8" φ

24" × 24" Module-Adjustable
Clg. Duffuser (Typ. This Dwg.)-
Adjust for 1000 CFM (Each)

Adjust (Each) for 500 CFM

2 (Each) 32,600 BTUH-Wall Mtd.
Air Conditioning Units to be
Installed by Owner to Provide
Total Design-Cooling Req'd.

PLAN

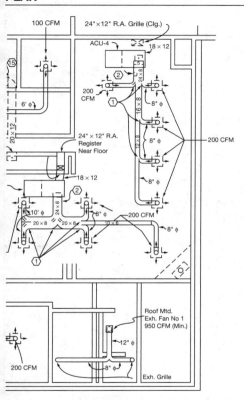

Notice that duct sizes and diffuser sizes are specified ACU = Air Conditioning Unit. This is a forced-air system, which is essentially the same for heating or air conditioning.

HEATING DETAIL

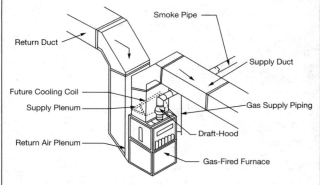

A perspective drawing of a basic gas forced-air heating system.

VAV CONTROL DIAGRAM

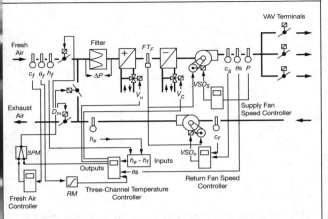

A control system that involves both pneumatic and electrical controls.

PILLOW BLOCK: PICTORAL VIEW

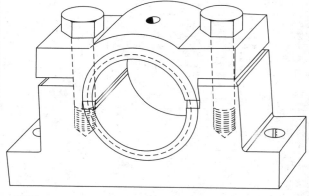

Pillow blocks are used to support motor shafts, making them a specialized type of bearing. Pillow blocks are important devices used with electric motors and associated mechanical devices.

PILLOW BLOCK DETAILS

Top View

End Elevation

Cast Iron

Brass

Side View

Tap   Drain

Oil Grooves

Top View

End View

Various views of a pillow block.

STEAM SYMBOLS

Relief Valve

Gate Valve

Vacuum Breaker

Pressure-Relief Valve

Motorized Valve

Governor

Thermometer

Pressure Gauge

Steam Trap

Steam Trap with
Integral Strainer

Balanced-Pressure
Thermostatic Trap

Float and Thermostatic
Steam Traps

Inverted Bucket Trap

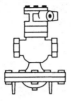

Control Valve with
Solenoid

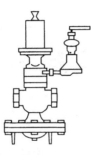

Pressure–Temperature
Regulator

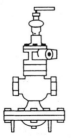

Temperature Regulator
with Solenoid

STEAM SYMBOLS (cont.)

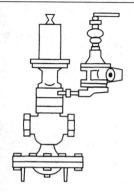

Pressure–Temperature
Regulator with Solenoid

H.C. Temperature
Regulator

H.C. Temperature
Regulator

Strainer

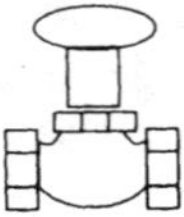

Balance Master Valve

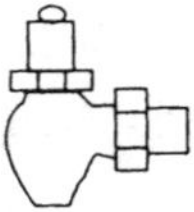

Balance Master Valve

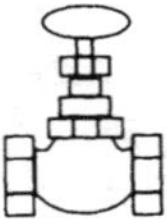

Radiator Valve

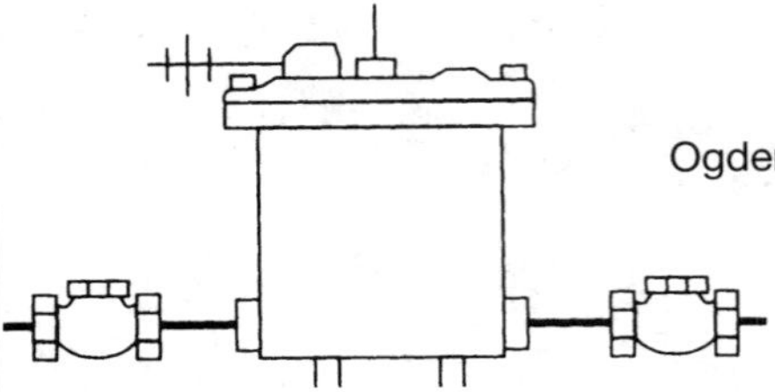

Ogden Pump

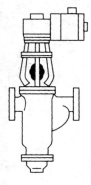

Scraper Strainer

Type VS Air Vent

Air Eliminator

PIPING SYMBOLS—VALVES, FITTINGS, AND SPECIALTIES

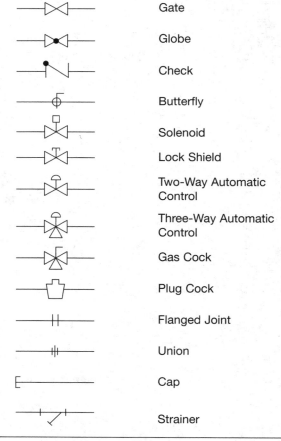

	Gate
	Globe
	Check
	Butterfly
	Solenoid
	Lock Shield
	Two-Way Automatic Control
	Three-Way Automatic Control
	Gas Cock
	Plug Cock
	Flanged Joint
	Union
	Cap
	Strainer

PIPING SYMBOLS—VALVES, FITTINGS, AND SPECIALTIES (cont.)

	Concentric Reducer
	Eccentric Reducer
	Pipe Guide
	Pipe Anchor
	Flow Direction
	Elbow Looking Up
	Elbow Looking Down
Up/Down	Pipe Pitch Up or Down
	Expansion Joint
	Expansion Loop
	Flexible Connection
T	Thermostat
	Thermostatic Trap

PIPING SYMBOLS—VALVES, FITTINGS, AND SPECIALTIES (cont.)

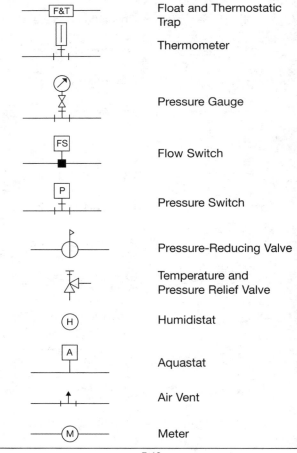

Float and Thermostatic Trap

Thermometer

Pressure Gauge

Flow Switch

Pressure Switch

Pressure-Reducing Valve

Temperature and Pressure Relief Valve

Humidistat

Aquastat

Air Vent

Meter

PIPING SYMBOLS—VALVES, FITTINGS, AND SPECIALTIES *(cont.)*

	Hose Bib
	Elbow
	Tee
	Y
	OS & Y Gate
	Shock Absorber
	House Trap
	P Trap
	Floor Drain
—IW—	Indirect Waste
— — -S— — —	Sanitary Below Grade
——S——	Sanitary Above Grade

PIPING SYMBOLS—VALVES, FITTINGS, AND SPECIALTIES *(cont.)*

— — —ST— — —	Storm Below Grade
——————ST——————	Storm Above Grade
— — — — — —	Vent
————CWV————	Combination Waste and Vent
— — —AW — — -	Acid Waste Below Grade
————AW————	Acid Waste Above Grade
— — - AV — — —	Acid Vent
———— _ _ <u>CW</u>	Cold Water
———— _ _ _ <u>HW</u>	Hot Water
——— _ _ _ _ <u>HWC</u>	Hot Water Circulation
————DWS————	Drinking Water Supply
————DWR————	Drinking Water Return

——G——	Gas—Low Pressure
——MG——	Gas—Medium Pressure
——HG——	Gas—High Pressure
——CA——	Compressed Air
——V——	Vacuum
——VC——	Vacuum Cleaning
——N——	Nitrogen
——N_2O——	Nitrous Oxide
——O——	Oxygen
——LOX——	Liquid Oxygen
——LPG——	Liquid Petroleum Gas

PIPING SYMBOLS—HEATING

—HPS—	High-Pressure Steam
—MPS—	Medium-Pressure Steam
—LPS—	Low-Pressure Steam
—HPR—	High-Pressure Return
—MPR—	Medium-Pressure Return
—LPR—	Low-Pressure Return
—BBD—	Boiler Blow-Down
—BD—	Boiler Blow-Off
—CP—	Condensate Pump Discharge
—VPD—	Vaccum Pump Discharge
—PPD—	Feedwater Pump Discharge
—MU—	Makeup Water
—V—	Air Relief Line (Vent)
—FOF—	Fuel Oil Flow
—FOR—	Fuel Oil Return
—FOS—	Fuel Oil Suction
—FOV—	Fuel Oil Tank Vent
—HWS—	Low-Temperature Hot Water Supply
—MTWS—	Medium-Temperature Hot Water Supply
—HTWS—	High-Temperature Hot Water Supply
—HWR—	Low-Temperature Hot Water Return
—MTWR—	Medium-Temperature Hot Water Return
—HTWR—	High-Temperature Hot Water Return

PIPING SYMBOLS—HEATING (cont.)

—A—	Compressed Air
—HWHS—	Hot Water Heating Supply
—HWHR—	Hot Water Heating, Return
—VAC—	Vacuum (Air)
—(NAME)E—	Existing Piping
–×–×–(NAME)–×–×–	Pipe to Be Removed

PIPING-SYMBOLS—AIR CONDITIONING AND REFRIGERATION

—RL—	Refrigerant Liquid
—RD—	Refrigerant Discharge
—RS—	Refrigerant Suction
—B—	Brine Supply
—BR—	Brine Return
—CWS—	Condenser Water Supply
—CWR—	Condenser Water Return
—CHWS—	Chilled Water Supply
—CHWR—	Chilled Water Return
—Fill—	Fill Line
—MW—	Makeup Water
—H—	Humidification Line
—D—	Drain
—B—	Brine Supply
—BR—	Brine Return

PIPING SYMBOLS—PLUMBING

————————	Soil, Waste, or Leader (above grade)
----------------	Soil, Waste, or Leader (below grade)

PIPING SYMBOLS—PLUMBING *(cont.)*

————————————	Vent
—·——·——·——·—	Cold Water
— — — — — —	Hot Water
— —·— —·— —·—	Hot Water Return
—G———————G—	Gas
———ACID———	Acid Water
——— DW ———	Drinking Water Flow
———DWR———	Drinking Water Return
——— VAC ———	Vacuum (Air)
——— A ———	Compressed Air
—— (NAME †) ——	Chemical Supply Pipes †

FIRE SAFETY DEVICES—SPRINKLER HEADS

—○——————○—	Upright
—●——————●—	Pendant
—⊗——————⊗—	Flush Mounted
▼ ▼	Sidewall

FIRE SAFETY DEVICES—SPRINKLER PIPING

——— S ———	Main Supplies
⊤	Control Valve
——— D ———	Drain
⊗——○——○— 4 in.	Riser and Branch (give size)
	Alarm Check Valve
	Dry Pipe Valve
	Pipe Hanger

FIRE SAFETY DEVICES—SIGNAL INITIATING DETECTORS

◔	Heat (Thermal)
◔	Smoke
▲	Gas
◔	Flame
▭ FCP	Control Panel

FIRE SAFETY DEVICES—VALVES FOR SELECTIVE ACTUATORS

—⊘—	Air Line
—|◯|—	Ball
—| |—	Butterfly
—⋈—	Diaphragm
—▷◁—	Gate
◁	Gate, Angle
—▶◀—	Globe
◀	Globe, Angle

FIRE SAFETY DEVICES—VALVES FOR SELECTIVE ACTUATORS *(cont.)*

	Globe, Stop Check
	Plug Valve
	Three Way

FIRE SAFETY DEVICES—VALVE ACTUATORS (MANUAL)

	Nonrising Stem
	Outside Stem and Yoke
	Lever
G	Gear

FIRE SAFETY DEVICES—ELECTRICAL

M	Motor
S	Solenoid

FIRE SAFETY DEVICES—PNEUMATIC VALVE ACTUATORS

A	Motor
A	Diaphragm

FIRE SAFETY DEVICES—PNEUMATIC VALVE ACTUATORS *(cont.)*

	Float
H	Hydraulic Piston

FIRE SAFETY DEVICES—VALVES (SPECIAL DUTY)

	Check, Swing Gate
	Check, Spring
	Control, Electric-Pneumatic
	Control, Pneumatic-Electric
	Hose End Drain
	Lock Shield
	Needle
	Pressure Reducing, Self-Contained
	Pressure Reducing, External Pressure

FIRE SAFETY DEVICES—VALVES
(SPECIAL DUTY) *(cont.)*

 Pressure Reducing, Differential Pressure

 Quick Opening

Quick Closing, Fusible Link

 Relief (R) or Safety (S)

 Solenoid

Square Head Cock

 Unclassified (number and specify) 1

FIRE PROTECTION PIPING SYMBOLS

———— F ———— Fire Protection Water Supply

———— WSP ———— Wet Standpipe

———— DSP ———— Dry Standpipe

———— CSP ———— Combination Standpipe

FIRE PROTECTION PIPING SYMBOLS (cont.)

——— SP ———	Automatic Fire Sprinkler
——○——○——	Upright Fire Sprinkler Heads
——●——●——	Pendant Fire Sprinkler Heads
	Fire Hydrant
	Wall Fire Department Connection
	Sidewalk Fire Department Connection
○—FHR—	Fire Hose Rack
FHC	Surface-Mounted Fire Hose Cabinet
FHC	Recessed Fire Hose Cabinet

FITTINGS

The following fittings are shown with screwed connections. The symbol for the body of a fitting is the same for all types of connections, unless otherwise specified. The types of connections are often specified for a range of pipe sizes, but are with the fitting symbol where required. For example, an elbow would be shown as:

	Flanged
	Soldered
	Welded
	Bell and Spigot
	Screwed
	Solvent Cement
	Bushing
	Cap
	Connection, Bottom
	Connection, Top

FITTINGS (cont.)

	Coupling (Joint)
	Cross
	Elbow, 90°
	Elbow, 45°
	Elbow, Turned Up
	Elbow, Turned Down
	Elbow, Reducing (show sizes)
	Elbow, Base
	Elbow, Long Radius
	Elbow, Double Branch
	Elbow, Side Outlet, Outlet Up
	Elbow, Side Outlet, Outlet Down
	Lateral
	Reducer, Concentric

	Reducer, Concentric, Straight Invert
	Reducer, Concentric, Straight Crown
	Tee
	Tee, Outlet Up
	Tee, Outlet Down
	Tee, Reducing (show sizes)
	Tee, Side Outlet, Outlet Up
	Tee, Side Outlet, Outlet Down
	Tee, Single Sweep
	Union

PIPING SPECIALTIES

	Air Eliminator
	Air Separator
	Alignment Guide

✕ PA	Anchor, Intermediate
⊠ PA	Anchor, Main
●	Ball Joint
▭ EJ-I	Expansion Joint
⊓	Expansion Loop
⧄⧄⧄⧄	Flexible Connector
☐ FD	Floor Drain
‖ OFM-I	Flowmeter, Orifice
▱ VFM-I	Flowmeter, Venturi
⊽ FS	Flow Switch
Y	Funnel Drain, Open
↓ H	Hanger, Rod
↓ H	Hanger, Spring
▭	Heat Exchanger, Liquid

PIPING SPECIALTIES *(cont.)*

▭ RAD-I	Heat Transfer Surface (indicate type)
P → R	Pitch of Pipe, Rise (R) Drop (D)
⊘H	Pressure Gauge and Cock
PS	Pressure Switch
⊘ CW-I	Pump (indicate use)
PSD	Pump Suction Diffuser
‖ ‖	Spool Piece, Flanged
	Strainer
	Strainer, Blow-off
⊗	Strainer, Duplex
▭ FO	Tank (indicate use)
⊡	Thermometer

	PIPING SPECIALTIES *(cont.)*
⊤	Thermometer Well, Only
T⟋	Thermostatic, Electirc
Ⓣ	Thermostatic, Pneumatic
Ⓣ	Thermostatic, Self-Contained
⊗F & T	Traps, Steam (indicate type)
∪	Trap, water
☐UH 1,2	Unit Heater (indicate type)

AIR-MOVING DEVICES AND COMPONENTS—FANS

⊕ R 1,2	Axial Flow
⊙ S 1, 2	Centrifugal
☐⊰ E 1, 2	Propeller
⊗ SVR - 1	Roof Ventilator, Intake
⊠ EVR - 1	Roof Ventilator, Exhaust
▣	Roof Ventilator, Louvered

AIR-MOVING DEVICES AND COMPONENTS—DUCTWORK

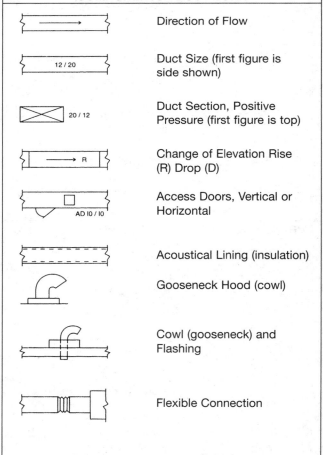

Direction of Flow

Duct Size (first figure is side shown)

12 / 20

Duct Section, Positive Pressure (first figure is top)

20 / 12

Change of Elevation Rise (R) Drop (D)

R

Access Doors, Vertical or Horizontal

AD 10 / 10

Acoustical Lining (insulation)

Gooseneck Hood (cowl)

Cowl (gooseneck) and Flashing

Flexible Connection

AIR-MOVING DEVICES AND COMPONENTS—DUCTWORK *(cont.)*

 Flexible Duct

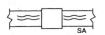 Sound Attenuator

 Terminal Unit, Mixing

 Terminal Unit, Reheat

  Terminal Unit, Variable Volume

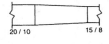 Transition

 Turning Vanes

 Detectors, Fire and/or Smoke

AIR-MOVING DEVICES AND COMPONENTS—DAMPERS

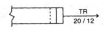 Adjustable Blank Off

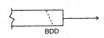 Back Draft Damper

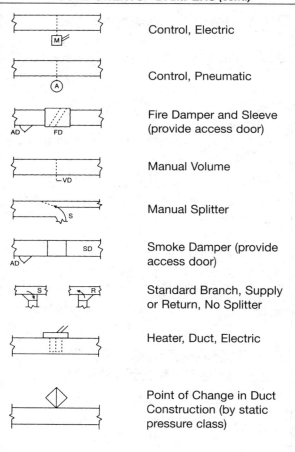

AIR-MOVING DEVICES AND COMPONENTS—DAMPERS *(cont.)*

Control, Electric

Control, Pneumatic

Fire Damper and Sleeve (provide access door)

Manual Volume

Manual Splitter

Smoke Damper (provide access door)

Standard Branch, Supply or Return, No Splitter

Heater, Duct, Electric

Point of Change in Duct Construction (by static pressure class)

20 x 12	Duct (first figure, side shown second figure, side not shown)
	Acoustical Lining Duct Dimensions for Net Free Area
	Direction of Flow
S 30 x 12	Duct Section (supply)
E OR R 20 x 12	Duct Section (exhaust or return)
R	Inclined Rise (R) or Drop (D) Arrow in Direction of Air Flow
	Transitions Give Sizes. Note F.O.T., Flat On Top, or F.O.B., Flat On Bottom, If Applicable

AIR-MOVING DEVICES AND
COMPONENTS—DAMPERS *(cont.)*

 Standard Branch for Supply and Return (no splitter)

Splitter Damper

 Volume Damper, Manual Operation

 Automatic Dampers, Motor Operated

 Access Door (AD) Access Panel (AP)

 Fire Damper:
SHOW ◀ VERTICAL POS.
SHOW ◆ HORIZ. POS.

 Ceiling Damper or Alternate Protection for Fire Rated CLG

Turning Waves

BDD Back Draft Damper

20 x 12 SG

700 CFM

Supply Grille (SG)

20 x 12 RG

700 CFM

Return (RG) or Exhaust (EG) Grille (note at FLR or GLG)

20 x 12 SR

700 CFM

Supply Register (SR) (A Grille + Integral Volume Control)

20 x 12 GR

700 CFM

Exhaust or Return Air Inlet Ceiling (indicate type)

20

700 CFM

Supply Outlet, Ceiling, Round (type as specified) (indicate flow direction)

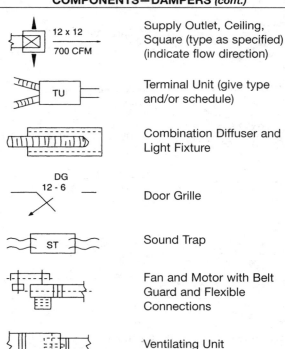

12 x 12 — 700 CFM	Supply Outlet, Ceiling, Square (type as specified) (indicate flow direction)
TU	Terminal Unit (give type and/or schedule)
	Combination Diffuser and Light Fixture
DG 12 - 6	Door Grille
ST	Sound Trap
	Fan and Motor with Belt Guard and Flexible Connections
	Ventilating Unit (type as specified)
	Unit Heater, Downblast

AIR-MOVING DEVICES AND COMPONENTS—DAMPERS *(cont.)*

 Unit Heater, Horizontal

 Unit Heater, Centrifugal Fan, Plan

 Thermostat

 Power or Gravity Roof Ventilator—Exhaust (ERV)

 Power or Gravity Roof Ventilator—Intake (SRV)

 Power or Gravity Roof Ventilator—Louvered

 Louvers and Screen

AIR-MOVING DEVICES AND COMPONENTS—GRILLES, REGISTERS, AND DIFFUSERS

 Supply Outlet

 Exhaust Inlet

TR 20 / 12 700	Grille or Register, Sidewall
CG 20 / 20 700	Grille or Register, Ceiling
40 / 36 700	Louver and Screen
20 / 12 L 200	Louver, Door or Wall
DG 12 x 6	Door Grille
UC 1/2" 100	Undercut Door
CD 300 20 / 12 300	Ceiling Diffuser, Rectangular
CD 20 NECK 1000	Ceiling Diffuser, Round
LD 1,2 48 / 3 300	Diffuser, Linear
100 100	Diffuser and Light Fixture Combination

AIR-MOVING DEVICES AND COMPONENTS—GRILLES, REGISTERS, AND DIFFUSERS *(cont.)*

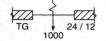 Transfer Grille Assembly

REFRIGERATION—COMPRESSORS

 Centrifugal

 Reciprocating

 Rotary

 Rotary Screw

REFRIGERATION—CONDENSERS

 Air Cooled

 Evaporate

 Water Cooled

REFRIGERATION—CONDENSING UNITS

 Air Cooled

 Water Cooled

REFRIGERATION—CONDENSER-EVAPORATOR

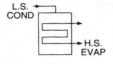 Cascade System

L.S. COND

H.S. EVAP

REFRIGERATION—COOLING TOWERS

 Cooling Tower

 Spray Pond

 Tank, Closed

 Tank, Open

REFRIGERATION—EVAPORATORS

 Finned Coil

Forced Convection

 Immersion Cooling Unit

 Plate Coil

 Pipe Coil

REFRIGERATION—LIQUID CHILLERS (CHILLERS ONLY)

 Direct Expansion

 Flooded

REFRIGERATION—CHILLING UNITS

 Absorption

 Centrifugal

REFRIGERATION—CHILLING UNITS *(cont.)*

 Reciprocating

 Rotary Screw

CONTROLS—REFRIGERATION

 Capillary Tube

 Expansion Valve, Hand

Expansion Valve, Automatic

 Expansion Valve, Thermostatic

 Float Valve, High Side
Float Valve, Low Side

 Thermal Bulb

 Solenoid Valve

 Constant-Pressure Valve, Suction

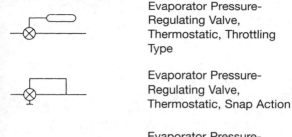

Evaporator Pressure-Regulating Valve, Thermostatic, Throttling Type

Evaporator Pressure-Regulating Valve, Thermostatic, Snap Action

Evaporator Pressure-Regulating Valve, Throttling Type, Evaporator Side

Compressor Suction Valve, Pressure Limiting, Throttling Type, Compressor Side

Thermo-Suction Valve

Snap Action Valve

Refrigerant Reversing Valve

CONTROLS—TEMPERATURE OR TEMPERATURE-ACTUATED ELECTRICAL OR FLOW CONTROLS

 Thermostat, Self-Contained

 Thermostat, Remote Bulb

CONTROLS PRESSURE OR PRESSURE-ACTUATED ELECTRICAL OR FLOW CONTROLS

 Pressure Switch

 Pressure Switch, Dual (High–Low)

 Pressure Switch, Differential Oil Pressure

 Automatic Reducing Valve

 Automatic Bypass Valve

 Valve, Pressure Reducing

 Valve, Condenser Water Regulating

AUXILLIARY EQUIPMENT—REFRIGERANTS

	Filter
	Strainer
	Filter and Drier
	Scale Trap
	Drier
	Vibration Absorber
	Heat Exchanger
	Oil Separator
	Sight Glass
	Fusible Plug
	Rupture Disc
	Receiver, High Pressure, Horizontal

AUXILLIARY EQUIPMENT — REFRIGERANTS *(cont.)*

Receiver, High Pressure, Vertical

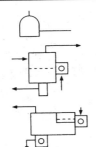

Receiver, Low Pressure

Intercooler

Intercooler/Desuperheater

ENERGY RECOVERY EQUIPMENT — CONDENSER, DOUBLE BUNDLE

Condenser, Double Bundle

AIR-TO-AIR ENERGY RECOVERY EQUIPMENT

Rotary Heat Wheel

Coil Loop

Heat Pipe

 Fixed Plate

 Plate Fin, Cross Flow

POWER RESOURCES

 Motor, Electric (number indicates horsepower)

MISCELLANEOUS PLAN SYMBOLS

◕	New Connection to Existing
ⓣ	Thermostat
T	Temperature Sensor
⒣	Humidity
H	Humidity Sensor
ⓢ	Switch

MISCELLANEOUS PLAN SYMBOLS *(cont.)*

| P | Pressure Sensor |

(x x) Sheet Note (number), Applies Only to the sheet it appears on

| x x | Coordination Point between Floor Plans and Diagrams (number) |

⟨ X ⟩ Demolition Note (number)

Ⓧ Ⓛ Plenum Light

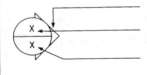

Direction of View

Section Number

Drawing on Which Section or Detail Is Shown

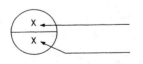

Detail or Section Number

Drawing from Which Section or Detail Is Taken

MISCELLANEOUS PLAN SYMBOLS (cont.)

Equipment Symbol (See Schedule)
Equipment Designation
Equipment Reference Number
System Number if Applicable

Equipment Symbol (See Schedule)
Equipment Designation
Equipment Reference Number

Existing to Remain

Existing to Be Removed

New Work

Supply Air Up

Supply Air Down

Exhaust or Return Air Up

Exhaust or Return Air Down

	Round Duct Up
	Round Duct Down
24 x 36	Rectangular Duct Size (first figure side shown)
36 x 14ø	Flat Oval Duct Size (first figure side shown)
→	Direction of Flow
UP DN	Duct Inclined Rise or Drop in Direction of Flow
	90° Elbow with Turning Vanes
	45° Elbow (no vanes)
	Supply or Return Branch Connection
	Supply or Return with Spin Collar Connection

MISCELLANEOUS PLAN SYMBOLS *(cont.)*

	Lateral Connection Round Ductwork
	Conical Tee Round Ductwork
	Duct with Internal Lining

CONTROL DIAGRAM SYMBOLS

AC INV	AC Inverter
AFS	Air Flow Station
AMS	Air Measuring Station
– – – – –	Control Tubing
— — —	Control Wiring
DO	Damper Operator
DPS	Differential Pressure Switch
DPT	Differential Pressure Transmitter

CONTROL DIAGRAM SYMBOLS *(cont.)*

| EP | Electric-Pneumatic Switch |

| ES | End Switch |

| FZ | Freezestat |

| FS | Flow Switch |

| FT | Flow Transmitter |

| H | Humidifier |

| —□— H | Humidity Sensor |

| —□— HL | High Limit Switch |

| HC | Humidity Controller |

| HOA | Hand-Off-Automatic Switch |

| MS | H O A | Combined Motor Starter and Hand-Off-Automatic Switch |

| M | Motor |

| MS | Motor Starter |

CONTROL DIAGRAM SYMBOLS *(cont.)*

| NC | Normally Closed |

| NO | Normally Open |

| PS | Pressure Sensor |

| PC | Pressure Controller |

| PE | Pneumatic-Electric Switch |

| PT | Pressure Transmitter |

| R | Relay |

| SDT | Smoke Detector |

| SP | Static Pressure with Pitot Tube |

| SPT | Static Pressure Transmitter |

| SR | Switching Relay |

| SS | Signal Selector |

CONTROL DIAGRAM SYMBOLS *(cont.)*

T	Temperature Sensor
⊂—T	Temperature Sensor with Bulb
TC	Temperature Controller
TDR	Time-Delay Relay
VSD	Variable-Speed Drive
▫—VP	Velocity Pressure Sensor
VC	Volume Controller
VT	Volume Transmitter
⊕	Reference Static Pressure
Ⓥ	Gauge
⊠	Light
⦿	On-Off Switch
⟨‾⟩	Three-Position Switch

CONTROL DIAGRAM SYMBOLS *(cont.)*

| ATC | Automatic Temperature Control Panel |

🖳   CRT Display

⊗   PC with Printer

Ⓜ   Main Air

Ⓗ   Humidistat

Ⓣ   Thermostat

BASIC WELDING SYMBOL

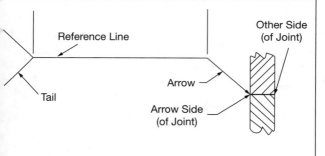

WELDING AND BRAZING SYMBOLS

			Groove				
Square	‖ Scarf	V	Bevel	U	J	Flare-V	Flare-Bevel
⊥⊥	⫽	⋁	⋁	⋃	⋃	⋋⋌	⋌⋌

Fillet	Plug or Slot	Spot or Projection	Seam	Back or Backing	Surfacing	Flange	
						Edge	Corner
◺	▭	◯	⊖	⌣	⌒⌒	⊔	⊔

AW

Tail section designates arc welding.

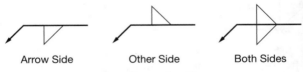

Arrow Side   Other Side   Both Sides

Locations of welds.

CHAPTER 8
Electrical Drawings

ELECTRICAL DRAWINGS

The electrical drawings cover the complete design and layout of the electrical wiring system for lighting, power, signals, communications, electronic systems, and fiber optics.

Electrical plans are prepared by certified professional electrical engineers, and each page should be sealed by the engineer who prepared them.

Electrical drawings may include a site plan showing the method of bringing power to the building, and floor plans showing power outlets, lighting fixtures, panelboards, and other items. A variety of schedules are included for panelboards, lighting fixtures, motors and other equipment. Specialized schematic, riser, and wiring diagrams may also be included.

As mentioned in Chapter 7, electrical control systems and related devices may be shown on the mechanical sheets of a set of blueprints. Electrical motors, disconnect switches, and motor starters may also be shown on these sheets.

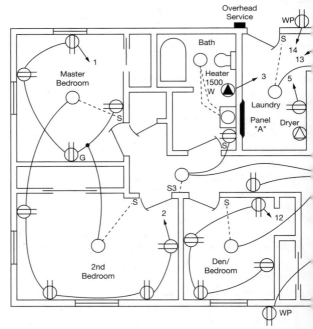

A typical electrical drawing for a house. The home runs (the circuit between the panel and first outlet) are shown with arrows, and circuit numbers are called out.

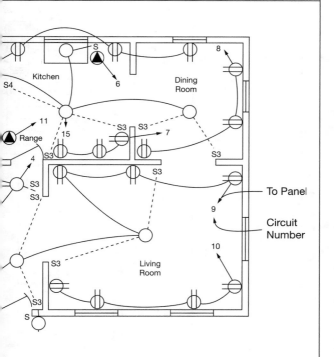

Kitchen

6

Dining
Room

8

S4

11

Range

15

7

4

To Panel

9

Circuit
Number

10

Living
Room

S

8-3

ELECTRICAL PLAN

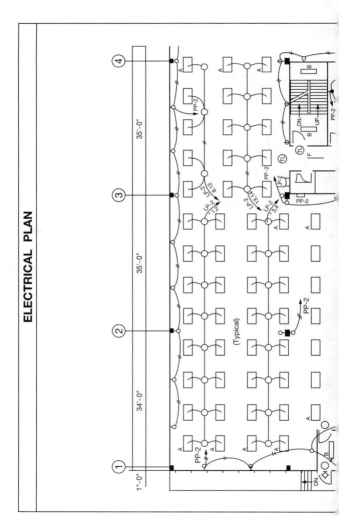

8-4

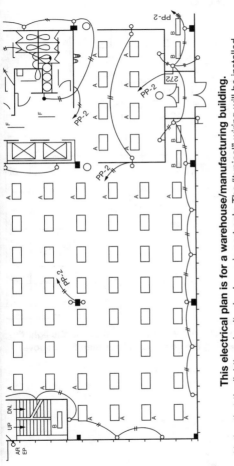

This electrical plan is for a warehouse/manufacturing building.

Notice that the lighting circuitry is shown in part only. The "typical" wiring will be installed throughout the building.

REFLECTED CEILING PLAN

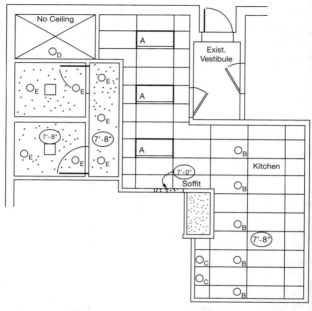

Reflected ceiling plans are very good for showing the light fixtures. You are viewing the ceiling and floor plan from above.

POWER RISER DIAGRAM

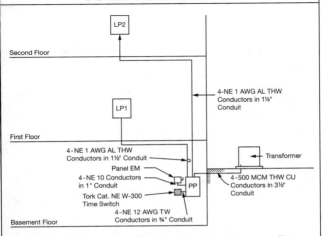

Power riser diagram for 400-A service feeding a small two-story commercial building.

ONE-LINE DIAGRAM

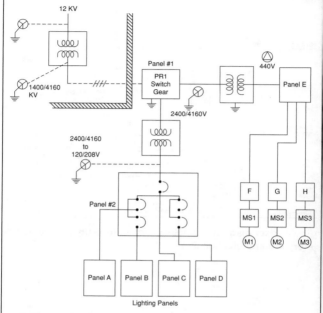

Schematic diagram of a power distribution system for a fairly large commercial facility.

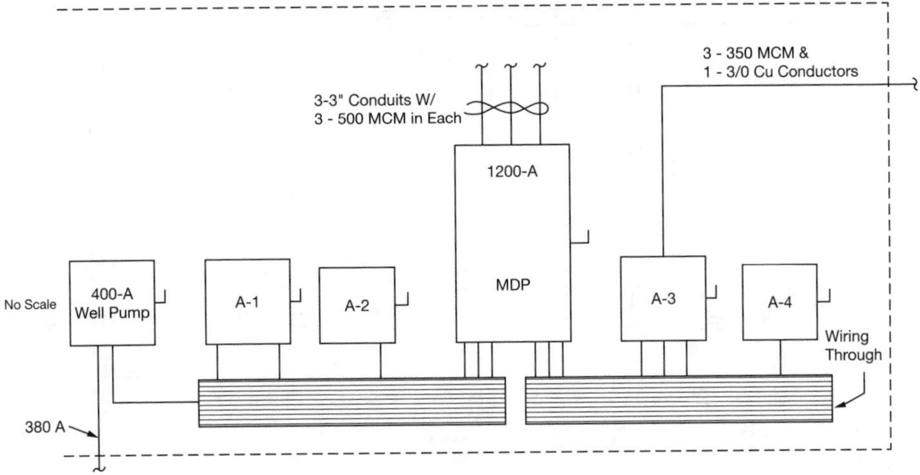

A block diagram, showing a neatly pictoral view of equipment in an electrical room.

3 - 350 MCM &
1 - 3/0 Cu Conductors

Wiring Through

A-4

A-3

1200-A

MDP

3-3" Conduits W/
3 - 500 MCM in Each

A-2

A-1

No Scale

400-A
Well Pump

380 A

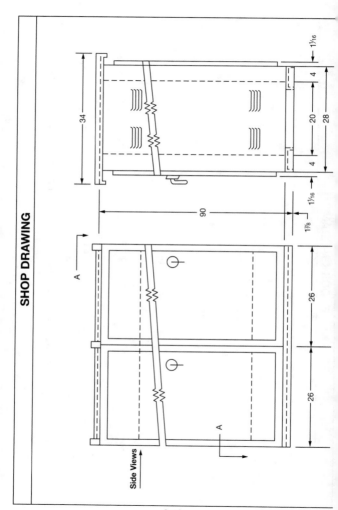

SHOP DRAWING

Side Views

8-10

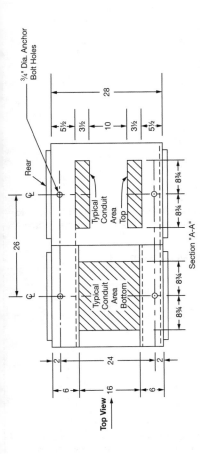

A detailed drawing of a motor control center. It is called a shop drawing because it is the type used in the workshop that manufactures the unit.

8-11

SCHEMATIC DIAGRAM

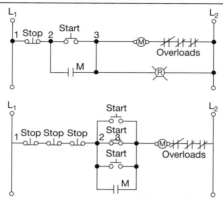

Schematic diagrams show components and their relation to each other *electrically* but not *physically*. These are motor starter circuits.

WIRING DIAGRAM

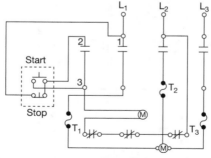

Wiring diagrams show components in their physical positions. This is a motor starter and is very similar to the schematic above.

TEMPORARY POLE

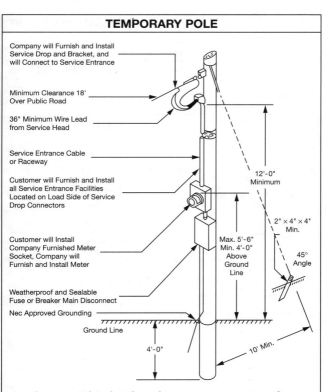

Company will Furnish and Install Service Drop and Bracket, and will Connect to Service Entrance

Minimum Clearance 18' Over Public Road

36" Minimum Wire Lead from Service Head

Service Entrance Cable or Raceway

Customer will Furnish and Install all Service Entrance Facilities Located on Load Side of Service Drop Connectors

Customer will Install Company Furnished Meter Socket, Company will Furnish and Install Meter

Weatherproof and Sealable Fuse or Breaker Main Disconnect

Nec Approved Grounding

Ground Line

12'-0" Minimum

2" × 4" × 4" Min.

45° Angle

Max. 5'-6" Min. 4'-0" Above Ground Line

4'-0"

10' Min.

A perspective drawing of a temporary power pole.

Panel A	Type NQO		Mounting Recess	
Use and/or Area Served		C/B	Cir. No	Load φ A
General-Den		20 / 1	1	
General-Living RM			3	
Lites-Front			5	
Receps-Planters			7	
Lites-Garage, Rear			9	
Receps-Garage			11	
Garage Door			13	
Spare			15	
Washer			17	
Freezer			19	
Future Site Lites		20 / 1	21	
Clothes Dryer		30	23	
		/ 2	25	
Water Heater		30	27	
		/ 2	29	
Relay Panel Power		20 / 1	31	
Garage Door Motor			33	
Spare			35	
Spare		20 / 1	37	
Space Only			39	
Space Only			41	

A typical branch circuit schedule.

8-14

SCHEDULE

	120/240V 1φ 3W			Mains 225 A
φ B	Cir. No	C/B		Use and/or Area Served
	2	20 / 1		General-B.R., Bath
	4			General-Sewing RM
	6			General-Mstr B.R.
	8			Lites-Util, Kitchen
	10			Furnace Fan
	12			Water Softener Cont.
	14			Kitchen Receps
	16			Kitchen Receps
	18			Disposer, D.W.
	20			Pool Lite
	22	20 / 1		Pool Pump
	24	50 /		Range
	26	2		
	28	40 /		Oven
	30	2		
	32	20 / 1		Spare
	34			Spare
	36			Spare
	38	20 / 1		Spare
	40			Space Only
	42			Space Only

LIGHTING FIXTURE SCHEDULE

Mark	Manufacture	Cat. No.	Wattage	Finish	Remarks
A	Lithonia	2G 440-A12	4-40	White	Grid Ceiling Acrylic.
B	Prescolite	4020	100	Black	
C	Prescolite	4452	4-60	Black	
D	Prescolite	7HV	75.R.30	White	
E	Prescolite	1252-916	75.R.30	Bronzotic	
F	Prescolite	540	100W	White	8" Tube
G		Recessed Ceiling Mt.	100W		Suitable for Use in Sauna
H	Lithonia	G 240-A12	2-40	White	Surface
J	Spaulding	832	60	Bronze Anod.	Wall Bracket
K	Prescolite	37D-3	2-60	Alum	6" A.F.F.
L	Prescolite	93020	175W-MV	Black	
N	Prescolite	4220	100	Black	
O	Allowance	of $ 250.00	Outlet		Outlet Only
P	———	———	100		Porcelain Socket
R	Prescolite	1171-900	2.50 R20	Bronzotic	

A lighting fixture schedule for an office building.

TELEPHONE RISER

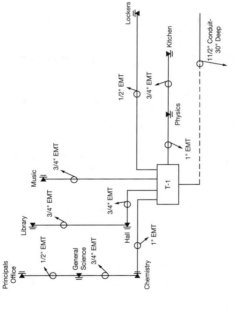

Schematic diagram of the raceways for a telephone system.

GENERAL ELECTRICAL PLAN SYMBOLS

(M)   Electric Motor

(G)   Electric Generator

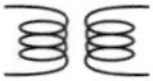

Power Transformer

Pothead (cable termination)

(WH)   Electric Watt-hour Meter

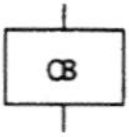

Circuit Element (e.g., circuit breaker)

Circuit Breaker

Fusible Element

Single-Throw Knife Switch

GENERAL ELECTRICAL PLAN SYMBOLS *(cont.)*

Double-Throw Knife Switch

Ground

Battery

LIGHTING OUTLET SYMBOLS

Ceiling   **Wall**

Surface or Pendant
Incandescent, Mercury Vapor,
or Similar Lamp Fixture

Recessed Incandescent, Mercury
Vapor, or Similar Lamp Fixture

Surface or Pendant
Individual Fluorescent Fixture

Recessed Individual
Fluorescent Fixture

LIGHTING OUTLET SYMBOLS *(cont.)*

Symbol	Description
`| O		|`	Surface or Pendant Continuous-Row Fluorescent Fixture
`| OR		|`	Recessed Continuous-Row Fluorescent Fixture*
(bare-lamp strip symbol)	Bare-Lamp Fluorescent Strip**

Ceiling **Wall**

Symbol	Description
(X) — (X)	Surface or Pendant Exit Light
(XR) — (XR)	Recessed Exit Light
(B) — (B)	Blanked Outlet
(J) — (J)	Junction Box
(L) — (L)	Outlet Controlled by Low-Voltage Switching When Relay is Installed in Outlet Box

---

* In the case of combination continuous-row fluorescent and incandescent spotlights, use combinations of the above standard symbols.

** In the case of a continuous-row bare-lamp fluorescent strip above an area-wide diffusion means, show each fixture run, using the standard symbol; indicate area of diffusing means and type of light shading and/or drawing notation.

RECEPTACLE OUTLET SYMBOLS

Single Receptacle Outlet

Duplex Receptacle Outlet

Triplex Receptacle Outlet

Quadruplex Receptacle Outlet

Duplex Receptacle
Outlet—Split Wired

Triplex Receptacle
Outlet—Split Wired

Single Special-Purpose
Receptacle Outlet*

Duplex Special-Purpose
Receptacle Outlet*

Range Outlet

Unless noted, assume every receptacle will be grounded, and will have a separate grounding contact.
* *Use numeral or letter, either within the symbol or as a subscript alongside the symbol, keyed to explanation in the drawing list of symbols, to indicate type of receptacle or usage.*

RECEPTACLE OUTLET SYMBOLS *(cont.)*

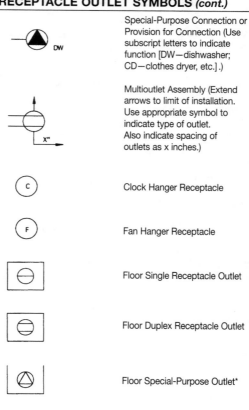

Special-Purpose Connection or Provision for Connection (Use subscript letters to indicate function [DW—dishwasher; CD—clothes dryer, etc.].)

Multioutlet Assembly (Extend arrows to limit of installation. Use appropriate symbol to indicate type of outlet. Also indicate spacing of outlets as x inches.)

Clock Hanger Receptacle

Fan Hanger Receptacle

Floor Single Receptacle Outlet

Floor Duplex Receptacle Outlet

Floor Special-Purpose Outlet*

Floor Telephone Outlet—Public

---

* Use numeral or letter, either within the symbol or as a subscript alongside the symbol, keyed to explanation in the drawing list of symbols, to indicate type of receptacle or usage.

Floor Telephone Outlet—Private

Not Part of the Standard: example of the use of several floor outlet symbols to identify a 2-, 3-, or more-gang floor outlet

Underfloor Duct and Junction Box for Triple, Double, or Single Duct System Indicated by Number of Parallel Lines

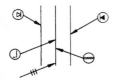

Not Part of the Standard: example of use of various symbols to identify location of different types of outlets or connections for underfloor duct or cellular floor systems

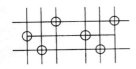

Cellular Floor Header Duct

SWITCH OUTLET SYMBOLS

S	Single-Pole Switch
S_2	Double-Pole Switch
S_3	Three-Way Switch
S_4	Four-Way Switch
S_K	Key-Operated Switch
S_P	Switch and Pilot Lamp
S_L	Switch for Low-Voltage Switching System
S_{LM}	Master Switch for Low-Voltage Switching System
$\ominus S$	Switch and Single Receptacle
$\ominus S$	Switch and Double Receptacle
S_D	Door Switch

S T	Time Switch
S CB	Circuit-Breaker Switch
S MC	Momentary Contact Switch or Pushbutton for Other than Signaling System

SIGNALING SYSTEM OUTLET SYMBOLS FOR INSTITUTIONAL, COMMERCIAL, AND INDUSTRIAL OCCUPANCIES

These symbols are recommended by the American Standards Association, but are not used universally. The reader should remember not to assume that these symbols will be used on any certain plan, but to always check the symbol list on the plans, and verify if these symbols are actually used.

Basic Symbol	Examples of Individual Item Identification with Description (Not Part of the Standard)
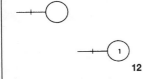	*Nurse Call System Devices (any type)* Nurses' Annunciator (can add a number after it as to indicate number of lamps) EX:12

SIGNALING SYSTEM OUTLET SYMBOLS FOR INSTITUTIONAL, COMMERCIAL, AND INDUSTRIAL OCCUPANCIES *(cont.)*

Basic Symbol	Examples of Individual Item Identification with Description (Not Part of the Standard)
─┤─(2)	Call Station, Single Cord, Pilot Light
─┤─(3)	Call Station, Double Cord, Microphone Speaker
─┤─(4)	Corridor Dome Light, One Lamp
─┤─(5)	Transformer
─┤─(6)	Any other item on same system—use numbers as required
─┤─◇	*Paging System Devices (any type)*
─┤─⟨ 1 ⟩	Keyboard
─┤─⟨ 2 ⟩	Flush Annunciator
─┤─⟨ 3 ⟩	Two-Face Annunciator
─┤─⟨ 4 ⟩	Any other item on same system—use numbers as required

SIGNALING SYSTEM OUTLET SYMBOLS FOR INSTITUTIONAL, COMMERCIAL, AND INDUSTRIAL OCCUPANCIES (cont.)

Basic Symbol	Examples of Individual Item Identification with Description (Not Part of the Standard)
	Fire Alarm System Devices (any type), Including Smoke and Sprinkler Alarm Devices
1	Control Panel
2	Station
3	10-inch Gong
4	Pre-signal Chime
5	Any other item on same system—use numbers as required
	Staff Register System Devices (any type)
1	Phone Operators' Register
2	Entrance Register—Flush
3	Staff Room Register

SIGNALING SYSTEM OUTLET SYMBOLS FOR INSTITUTIONAL, COMMERCIAL, AND INDUSTRIAL OCCUPANCIES *(cont.)*

Basic Symbol	Examples of Individual Item Identification with Description (Not Part of the Standard)
◇4	Transformer
◇5	Any other item on same system—use numbers as required
⬡	*Electric Clock System Devices (any type)*
⬡1	Master Clock
⬡2	12-inch Secondary—Flush
⬡3	12-inch Double Dial—Wall Mounted
⬡4	18-inch Skeleton Dial
⬡5	Any other item on same system—use numbers as required
◁	*Public Telephone System Devices*
◁1	Switchboard

8-28

Basic Symbol	Examples of Individual Item Identification with Description (Not Part of the Standard)

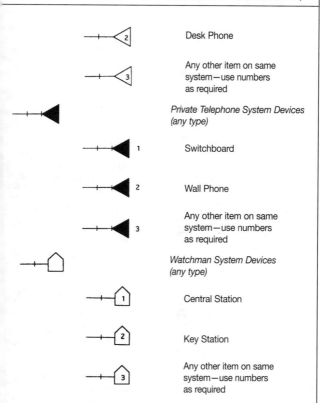

Desk Phone

Any other item on same system—use numbers as required

Private Telephone System Devices (any type)

Switchboard

Wall Phone

Any other item on same system—use numbers as required

Watchman System Devices (any type)

Central Station

Key Station

Any other item on same system—use numbers as required

SIGNALING SYSTEM OUTLET SYMBOLS FOR INSTITUTIONAL, COMMERCIAL, AND INDUSTRIAL OCCUPANCIES (cont.)

Basic Symbol	Examples of Individual Item Identification with Description (Not Part of the Standard)	
⊸◁		*Sound System*
	⊸◁1	Amplifier
	⊸◁2	Microphone
	⊸◁3	Interior Speaker
	⊸◁4	Exterior Speaker
	⊸◁5	Any other item on same system—use numbers as required
⊸◯		*Other Signal System Devices*
	⊸①1	Buzzer
	⊸②2	Bell
	⊸③3	Pushbutton
	⊸④4	Annunciator
	⊸⑤5	Any other item on same system—use numbers as required

SIGNALING SYSTEM OUTLET SYMBOLS
FOR RESIDENTIAL OCCUPANCIES

When a descriptive symbol list is not employed, use the following signaling system symbols to identify standardized, residential-type, signal-system items on residential drawings.

▪	Pushbutton
◁	Buzzer
◁	Bell
◁	Combination Bell-Buzzer
CH	Chime
◇	Annunciator
D	Electric Door Opener
M	Maid's Signal Plug
☐	Interconnection Box
BT	Bell-Ringing Transformer
▶	Outside Telephone
▷	Interconnecting Telephone
R	Radio Outlet
TV	Television Outlet

PANELBOARDS, SWITCHBOARDS, AND RELATED EQUIPMENT SYMBOLS

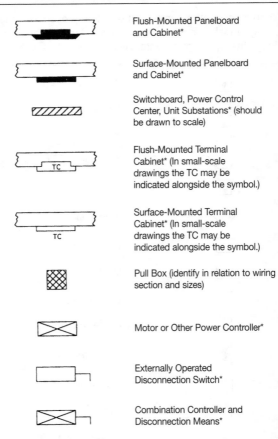

Flush-Mounted Panelboard and Cabinet*

Surface-Mounted Panelboard and Cabinet*

Switchboard, Power Control Center, Unit Substations* (should be drawn to scale)

Flush-Mounted Terminal Cabinet* (In small-scale drawings the TC may be indicated alongside the symbol.)

Surface-Mounted Terminal Cabinet* (In small-scale drawings the TC may be indicated alongside the symbol.)

Pull Box (identify in relation to wiring section and sizes)

Motor or Other Power Controller*

Externally Operated Disconnection Switch*

Combination Controller and Disconnection Means*

*Identify by notation or schedule.

BUS DUCTS AND WIREWAYS SYMBOLS

| T | T | T | Trolley duct* |

| B | B | B | Busway (service, feeder, or plug-in)* |

| C | C | C | Cable trough ladder or channel* |

| W | W | W | Wireway* |

REMOTE CONTROL STATION SYMBOLS FOR MOTORS OR OTHER EQUIPMENT

Pushbutton Station

| F | Float Switch—Mechanical |

| L | Limit Switch—Mechanical |

| P | Pneumatic Switch—Mechanical |

Electric Eye—Beam Source

Electric Eye—Relay

(T) Thermostat

Identify by notation or schedule.

CIRCUITING SYMBOLS

Wiring method identification by notation on drawing or in specification.

————————————	Wiring Concealed in Ceiling or Wall
– – – – – – – – – –	Wiring Concealed in Floor
– – – – – – – – – –	Wiring Exposed *Note: Use heavyweight line to identify service and feeders. Indicate empty conduit by notation CO (conduit only).*
2 1 ————————→	Branch-Circuit Home Run to Panelboard Number of arrows indicates number of circuits. (A numeral at each arrow may be used to identify circuit number.) *Note: Any circuit without further identification indicates a two-wire circuit. For more wires, indicate with cross lines.*
3 wires ——///——	Unless indicated otherwise, the wire size of the circuit is the minimum size required by the specification. Identify different functions of wiring system, e.g., signaling system, by notation or other means.
4 wires ——////——	
———————○	Wiring Turned Up
———————●	Wiring Turned Down

8-34

ELECTRIC DISTRIBUTION OR
LIGHTING SYSTEM—UNDERGROUND

| M | Manhole* |

| H | Handhole* |

| TM | Transformer Manhole or Vault* |

| TP | Transformer Pad* |

– – – – – – –   Underground Direct Burial Cable (Indicate type, size, and number of conductors by notation or schedule.)

Underground Duct Line (Indicate type, size, and number of ducts by cross-section identification of each run by notation or schedule. Indicate type, size, and number of conductors by notation or schedule.)

Streetlight Standard Feed from Underground Circuit*

Identify by notation or schedule.

ELECTRIC DISTRIBUTION OR LIGHTING SYSTEM—AERIAL

◯	Pole*
☼	Streetlight and Bracket*
△	Transformer*
————	Primary Circuit*
- - - - - - -	Secondary Circuit*
◗	Down Guy
—●—	Head Guy
—○—→	Sidewalk Guy
◖	Service Weather Head*

* Identify by notation or schedule.

ARRESTER, LIGHTING ARRESTER GAP SYMBOLS (ELECTRIC SURGE)

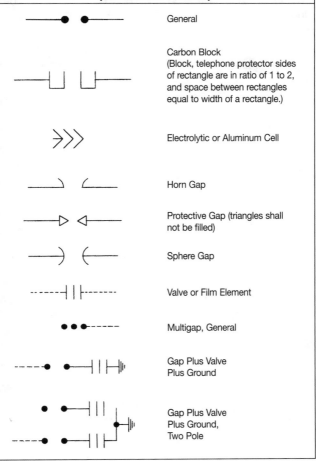

General

Carbon Block
(Block, telephone protector sides of rectangle are in ratio of 1 to 2, and space between rectangles equal to width of a rectangle.)

Electrolytic or Aluminum Cell

Horn Gap

Protective Gap (triangles shall not be filled)

Sphere Gap

Valve or Film Element

Multigap, General

Gap Plus Valve
Plus Ground

Gap Plus Valve
Plus Ground,
Two Pole

BATTERY SYMBOLS

The long line is always positive, but polarity may also be indicated as shown in the direct-current source.

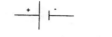 Generalized Direct-Current Source

One Cell

Multicell

Multicell Battery with Three Taps

 Multicell Battery with Adjustable Tap

CIRCUIT BREAKER SYMBOLS

If it is desired to show the condition that will cause the breaker to trip, a relay protective function symbol may be used alongside the break symbol. On a power diagram, the symbol may be used without other identification. On a composite drawing, the identifying letter CB may be added inside or adjacent to the square.

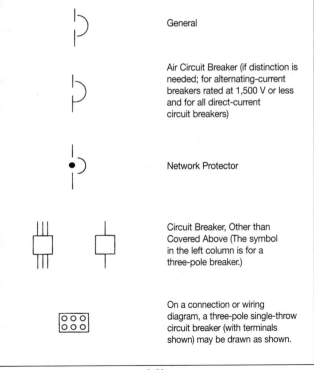

General

Air Circuit Breaker (if distinction is needed; for alternating-current breakers rated at 1,500 V or less and for all direct-current circuit breakers)

Network Protector

Circuit Breaker, Other than Covered Above (The symbol in the left column is for a three-pole breaker.)

On a connection or wiring diagram, a three-pole single-throw circuit breaker (with terminals shown) may be drawn as shown.

CIRCUIT BREAKER APPLICATION SYMBOLS

Three-Pole Circuit Breaker

Three-Pole Circuit Breaker with Thermal Overload Device in all Three Poles

Three-Pole Circuit Breaker with Magnetic Overload Device in All Three Poles

Three-Pole Circuit Breaker, Drawout Type

CIRCUIT RETURN SYMBOLS

Ground

> (a) A direct conducting connection to the earth or body of water

> (b) A conducting connection to a structure similar to an earth ground (frame of an air, space, or vehicle that is not conductively connected to earth)

Chassis or Frame Connection

> A conducting connection to a chassis or frame of a unit. The chassis or frame may be a substantial potential with respect to the earth or structure in which this chassis or frame is mounted.

Common Connections

> Conducting connections made to one another. All like-designated points are connected. *The asterisk is not part of the symbol. Identifying valves, letters, numbers, or marks replace the asterisk.

GENERATOR AND MOTOR SYMBOLS

◯	Basic
(GEN)	Generator
(MOT)	Motor
	Rotating Armature with Commutator and Brushes
	Compensating or Commutating
	Series
	Shunt, or Separately Excited Magnet, Permanent

WINDING SYMBOLS

Generator and motor winding symbols may be shown in the basic circle using the following representation.

	One-phase
⊗	Two-phase

	WINDING SYMBOLS *(cont.)*
	Three-Phase Wye—Ungrounded
	Three-Phase Wye—Grounded
	Three-Phase Delta
	Six-Phase Diametrical
	Six-Phase Double Delta

	APPLICATIONS FOR DIRECT CURRENT	
		Separately Excited Direct-Current Generator or Motor
		Separately excited Direct-Current Generator or Motor, with Commutating or Compensating Field Winding or Both
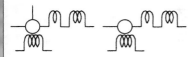		Compositely Excited Direct-Current Generator or Motor, with Commutating or Compensating Field Winding or Both

APPLICATIONS FOR DIRECT CURRENT *(cont.)*

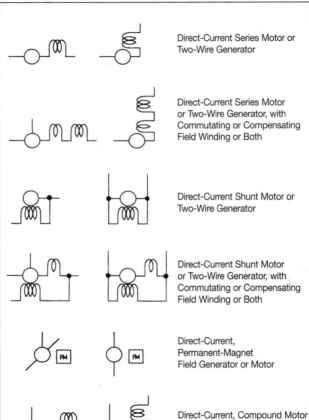

Direct-Current Series Motor or Two-Wire Generator

Direct-Current Series Motor or Two-Wire Generator, with Commutating or Compensating Field Winding or Both

Direct-Current Shunt Motor or Two-Wire Generator

Direct-Current Shunt Motor or Two-Wire Generator, with Commutating or Compensating Field Winding or Both

Direct-Current, Permanent-Magnet Field Generator or Motor

Direct-Current, Compound Motor or Two-Wire Generator or Stabilized Shunt Motor

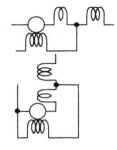

Direct-Current Compound Motor or Two-Wire Generator Or Stabilized Shunt Motor; with Commutating or Compensating Field Winding or Both

Direct-Current, Three-Wire Shunt Generator

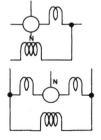

Direct-Current, Three-Wire Shunt Generator, with Commutating or Compensating Field Winding or Both

APPLICATIONS FOR DIRECT CURRENT *(cont.)*

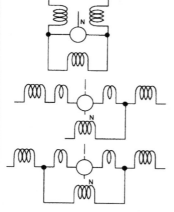

Direct-Current, Three-Wire
Compound Generator*

Direct-Current, Three-Wire
Compound Generator, with
Commutating or Compensating
Field Winding or Both*

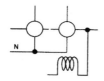

Direct-Current Balancer,
Shunt Wound*

---

* The broken line — - — indicates where line connection to a symbol is made and is not a part
of the symbol.

APPLICATIONS FOR DIRECT CURRENT *(cont.)*

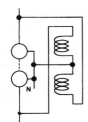

Direct-Current Balancer,
Shunt Wound*

ALTERNATING-CURRENT MOTOR SYMBOLS

Squirrel-Cage Induction Motor
or Generator, Split-Phase
Induction Motor or Generator,
Rotary Phase Converter or
Repulsion Motor

Wound-Rotor Induction Motor,
Synchronous Induction Motor,
Induction Generator, or Induction
Frequency Converter

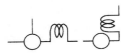

Alternating-Current Series Motor*

---

*The broken line — - — indicates where line connection to a symbol is made and is not a part
of the symbol.*

PATH—TRANSMISSION, CONDUCTION, CABLE, WIRING, ETC.

The entire group of conductors, or the transmission path required to guide the power or symbol, is shown by a single line. In coaxial and waveguide work, the recognition symbol is employed at the beginning and end of each type of transmission path as well as at intermediate points to clarify a potentially confusing diagram. For waveguide work, the mode may be indicated as well.

————————————	General; Conductive, Conductor or Wire Guided Conductor or Wire Path
—//— ═══════	Two Conductors or Conductive Paths of Wires
—///— ═══════	Three Conductors or Conductive Paths of Wires
———n/———	n Conductors or Conductive Paths of Wires
+	Crossing of Paths or Conductors Not Connected (The crossing is not necessarily at a 90° angle.)

JUNCTION OF PATHS/CONDUCTORS SYMBOLS

Junction of Paths or Conductors

Application: Junction of Paths, Conductor, or Cable (path type or size may be indicated on diagram)

Splice

Application: Splice of Same Size Cables

Junction of Conductors of Same Size or Different Size Cables (sizes of conductors should be indicated on diagram)

Junction of Connected Paths, Conductors, or Wires

POLARITY SYMBOLS

$+$ Positive

$-$ Negative

SWITCH SYMBOLS

Switch symbols are constructed of basic symbols for mechanical connections, contacts, etc., and normally a switch is represented in the de-energized position for switches having two or more positions where no operating force is applied. When actuated by a mechanical force, the functioning point is described by a clarifying note. Where switch symbols are in closed position, the terminals should be included for clarity.

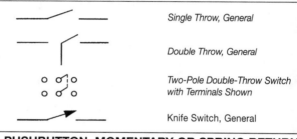

Single Throw, General

Double Throw, General

Two-Pole Double-Throw Switch with Terminals Shown

Knife Switch, General

PUSHBUTTON, MOMENTARY OR SPRING RETURN

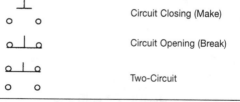

Circuit Closing (Make)

Circuit Opening (Break)

Two-Circuit

PUSHBUTTON, MAINTAINED OR NOT SPRING RETURN

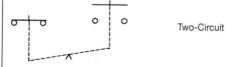

Two-Circuit

TRANSFORMER SYMBOLS

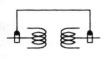
Transformer with Direct-Current Connections and Mode Suppression between Two Rectangular Waveguides

Transformer with Magnetic Core Shown

Shielded Transformer with Magnetic Core Shown

Transformer with Magnetic Core Shown and with a Shield between Windings (shield shown connected to the frame)

With Taps, One-Phase

Autotransformer, One-Phase

Adjustable

One-Phase, Two-Winding Transformer

TRANSFORMER SYMBOLS *(cont.)*

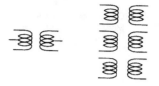 Three-Phase Bank of One-Phase, Two-Winding Transformer

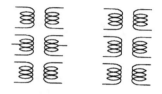 Polyphase Transformer

 Current Transformer(s)

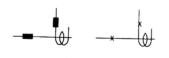 Current Transformer with Polarity Marking. (Instantaneous direction of current into one polarity mark corresponds to current out of the other polarity mark.)

TRANSFORMER SYMBOLS (cont.)

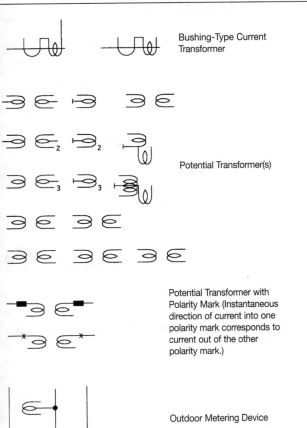

Bushing-Type Current Transformer

Potential Transformer(s)

Potential Transformer with Polarity Mark (Instantaneous direction of current into one polarity mark corresponds to current out of the other polarity mark.)

Outdoor Metering Device

TRANSFORMER WINDING CONNECTION SYMBOLS

Two-Phase Three-Wire

Two-Phase Three-Wire, Grounded

Two-Phase Four-Wire

Two-Phase Five-Wire, Grounded

Three-Phase Three-Wire, Delta or Mesh

Three-Phase Three-Wire, Delta, Grounded

Three-Phase Four-Wire, Delta

Three-Phase Four-Wire, Delta, Grounded

TRANSFORMER WINDING
CONNECTION SYMBOLS *(cont.)*

Three-Phase, Open Delta

Three-Phase, Open Delta,
Grounded at Common Point

Three-Phase, Open-Delta,
Grounded at Middle Point
of One Transformer

Three-Phase, Broken Delta

Three-Phase, Wye or Star

Three-Phase, Wye, Grounded Neutral

Three-Phase Four-Wire

MOTOR CONTROL SYMBOLS

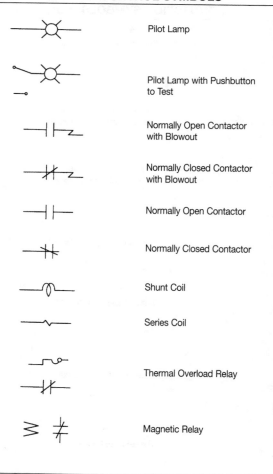

Pilot Lamp

Pilot Lamp with Pushbutton
to Test

Normally Open Contactor
with Blowout

Normally Closed Contactor
with Blowout

Normally Open Contactor

Normally Closed Contactor

Shunt Coil

Series Coil

Thermal Overload Relay

Magnetic Relay

MOTOR CONTROL SYMBOLS *(cont.)*

Limit Switch, Normally Open

Limit Switch, Normally Closed

Footswitch, Normally Open

Footswitch, Normally Closed

Vacuum Switch, Normally Open

Vacuum Switch, Normally Closed

Liquid-Level Switch, Normally Open

Liquid-Level Switch, Normally Closed

Temperature-Actuated Switch, Normally Open

Temperature-Actuated Switch, Normally Closed

Flow Switch, Normally Open

Flow Switch, Normally Closed

MOTOR CONTROL SYMBOLS (cont.)

	Momentary-Contact Switch, Normally Open
	Momentary-Contact Switch, Normally Closed
	Iron-Core Inductor
	Air-Core Inductor
	Single-Phase AC Motor
	Three-Phase, Squirrel-Cage Motor
	Two-Phase, Four-Wire Motor
	Wound-Rotor, Three-Phase Motor
	Armature
	Crossed Wires, Not Connected
	Crossed Wires, Connected
	Fuse

CHAPTER 9
Glossary

GLOSSARY

A

ABS: Plastic pipe used for plumbing construction.

Absolute Pressure: The total pressure measured from absolute vacuum; the sum of gauge optional and atmospheric pressure corresponding to the barometer reading expressed in PSI.

Absolute Zero: A point of total absence of heat, $-273.18°C$.

Absorption: Loss of power in an optical fiber, resulting from conversion of optical power into heat.

Abut: Joining end to end.

Accelerator: An additive used to speed curing time of freshly poured concrete.

Access Line: A line or circuit that connects a customer site to a network switching center or local exchange. Local loop.

Accessible: Having access, which may require removal of an access panel, etc.

Accumulator: A container in which fluid or gas is stored under pressure.

AC Current: Electrical current that reverses direction at regular intervals (cycles).

Acid Vent: A pipe venting an acid waste system.

Acid Waste: A pipe which conveys liquid waste matter.

AC Line Filter: Absorbs electrical interference.

Acme Thread: A thread used for feed screws.

Acoustical: Referring to the study of sound transmission.

Active Sludge: Sewage sediment, rich in destructive bacteria, that can be used to break down fresh sewage.

AC Voltage: Electrical pressure that reverses direction at regular intervals.

Adapter Fitting: Designed to fit two pipes or fittings different in design or material when connecting the two together would not be possible.

Adhesive: A bonding material.

Adjacent: Touching; next to.

Administrative Authority: The individual agency authorized by a political subdivision to enforce the provisions of a building code.

Admixtures: Materials added to concrete or mortar to alter it in some way, as an accelerator, retarder, etc.

Aeration: An artificial method in which water and air are brought into direct contact with each other.

Aerial Cable: Telecommunications cable installed on aerial supporting structures such as poles, sides of buildings, and other structures.

Aerobic: Bacteria living or active only in free oxygen.

Aggregate: Grades of sand, vermiculite, perlite, or gravel added to cement for concrete or plaster.

Air Break: A physical separation in which a drain discharges indirectly into a fixture, receptacle, or interceptor at a point below the rim of the receptacle to prevent backflow.

Air Chamber: A continuation of the water piping beyond the branch to fixtures finished with a cap designed to eliminate shock or vibration.

Air, Compressed: Air at any pressure greater than atmospheric pressure.

Air Drying: Method of removing excess moisture from lumber using natural circulation of air.

Air, Free: Air which is subject only to atmospheric conditions.

Air Gap: The unobstructed vertical distance through the free atmosphere between the lowest opening from any pipe or faucet conveying water or waste to a tank, plumbing fixture, receptor, or other device, and the flood level rim of the receptacle normally twice the diameter of the inlet.

Air-Handling Unit: A mechanical unit used for air conditioning or movement of air, as supply or exhaust.

Air, Standard: Air having a temperature of 70°F (21.1°C), a standard density of 0.0075 lb/ft (0.11 kg/m), and under pressure of 14.70 psia (101.4 kPa); the gas industry standard is 60°F (15.6°C).

Air Test: A test applied to a plumbing system upon its completion.

Alarm Check Valve: A check valve equipped with a signaling device which will annunciate a remote alarm.

Allowable Load: Maximum supportable load of any construction components(s).

GLOSSARY (cont.)

Allowable Span: Maximum length permissible for any framing component without support.

Alloy: A substance composed of two or more metals.

Alloy Pipe: A steel pipe with one or more elements other than carbon which give it greater resistance to corrosion and more strength.

Alternator: Converts mechanical energy into electrical energy.

Ambient Temperature: The prevailing temperature of the area surrounding an object.

American National Standards Institute (ANSI): A private organization that coordinates standards.

American Standard Code for Information Interchange (ASCII): A standard character set that (typically) assigns a 7-bit sequence to each letter, number, and selected control character.

American Standard Pipe Thread: A type of screw thread commonly used on pipe and fittings.

American Wire Gauge (AWG): Standard used to describe the size of a wire.

Ammeter: An instrument (meter) for measuring electrical current.

Ampacity: The amount of current (amps) that a conductor can carry without overheating.

Ampere (A): Unit of current measurement.

Ampere-hour: The flow of one ampere for one hour.

Amplitude: The size, in voltage, of signals in a data transmission.

Anaerobic: Bacteria living in the absence of free origin.

Anchor: A device used to secure pipes to a structure.

Anchor Bolt: A J- or L-shaped steel rod threaded on one end for securing structural members to concrete.

Angle of Bend: In a pipe, the angle between radial lines from the beginning and end of the bend to the center.

Angle Valve: A device in which the inlet and outlet are at right angles.

Annunciator: A sound-generating device that intercepts and speaks the condition of circuits.

Anode: The positive electrode in a battery.

GLOSSARY (cont.)

Anodize: An electrolytic means of coating aluminum or magnesium by oxidizing.

Apparent Power (P$_A$): Product of the voltage and current in a circuit calculated without considering the phase shift. Expressed in terms of volt-amperes (VA).

Approved: Accepted under an applicable specification or standard by the administrative authority.

Approved Ground: A grounding bus or strap suitable for connecting to data com equip.

Approved Testing Agency: An organization established for purposes of testing to approved standards.

Apron: A piece of window trim beneath the window sill; also used to designate the front of a building, such as the concrete apron in front of a garage.

Arbor: An axle on which a cutting tool is mounted.

Arc Tube: The light-producing element of an HID lamp.

Architect's Scale: A rule with scales indicating feet, inches, and fractions of inches.

Arcing: A luminous discharge formed by the span of electrical current across a space between terminals.

Area Drain: A receptacle designed to collect surface or rain water.

Areaway: The open space around foundation walls, doorways or windows to permit light and air to reach the below-ground-level floors.

Arrestor (Lightning): A device that reduces the voltage of a surge applied to its terminals and restores itself to its original condition.

Arterial Vent: A vent serving a drain and a public sewer.

Asbestos: A mineral material used for exterior wall siding and for fireproofing.

Ashlar: A stone cut by sawing to a rectangular shape.

Asphalt: A black material produced as a by-product of oil or coal.

Asphalt Shingle: A composition-type shingle used on a roof and is fire-resistant.

Aspirator: A fitting or device supplied with water or other fluid under positive pressure which passes through an integral orifice or constriction causing a vacuum.

Atmospheric Vacuum Breaker: A mechanical device consisting of a check valve opening to the atmosphere when the pressure in the piping drops to atmospheric.

Attenuation: Denotes the loss in strength of power between that transmitted and that received. Expressed as a ratio in decibels (dB).

Authority Having Jurisdiction: The organization responsible for approving equipment, installation, or procedure.

Autotransformer: Changes voltage level using the same common coil for both the primary and the secondary.

Awl: A tool used to mark wood or make pilot holes.

Awning Window: A window that is hinged at the top and the bottom swings outward.

B

Backboard: A wooden (or metal) panel used for mounting equipment.

Backbone: The main connectivity device of a distributed system.

Back Electromagnetic Force: The voltage created in an inductive circuit by a changing current flowing through the circuit.

Backfill: Any deleterious material (sand, gravel, etc.) used to fill an excavation.

Backflow: The flow of water or other liquids into the distributing pipes of a potable water supply from any source other than its intended source.

Backflow Connection: A condition in any arrangement where backflow may occur.

Backflow Preventer: Device to prevent backflow into the potable water system.

Backhoe: Self-powered excavation equipment.

Backing Ring: A metal strip used to prevent melted metal from entering a pipe when making a butt-welded joint.

Back-Siphonage: The flowing back of used or contaminated water from a fixture or vessel into a water supply pipe due to negative pressure in the pipe.

Backsplash: The vertical part of a countertop that runs along the wall to prevent splashing the wall.

Back Up: A condition in which waste water may flow back into another fixture or compartment but not into the potable water system.

GLOSSARY *(cont.)*

Backwater Valve: A device which permits drainage in one direction but has a check valve that closes against back pressure.

Baffle Plate: A tray or partition placed in process equipment to change the direction of flow.

Ballast: A component used with fluorescent lamps to provide the voltage necessary to strike the mercury arc, then limit the amount of current that flows through the lamp.

Ball Check Valve: A device used to stop flow in one direction while allowing flow in the opposite direction.

Balloon Framing: Wall construction extending from the foundation to the roof structure without interruption.

Ball Valve: A valve providing a tight shutoff.

Baluster: That part of a staircase which supports the handrail or bannister.

Balustrade: A complete handrail assembly.

Bank: An assemblage of fixed contacts.

Bank Plugs: Pieces of lumber driven into the ground so surveyors can string a line between them to measure grade.

Bannister: That part of a staircase which fits on top of the balusters.

Base: The lowest portion or lowest point of a stack of vertical pipe.

Baseboard: Molding covering the joint between a finished wall and the floor.

Base Shoe: A molding added at the bottom of a baseboard used to cover the edge of finish flooring or carpeting.

Batten: A narrow piece of wood used to cover a joint.

Batter Boards: Boards used to frame in the corners of a proposed building during layout and excavation.

Battery: A device that converts chemical energy into electrical current.

Battery of Fixtures: Two or more similar adjacent fixtures which discharge into a common horizontal waste or soil branch.

Batt Insulation: An insulating material to be installed between framing members.

Beam: A horizontal framing member made of steel or wood at least 5 inches thick and at least 7 inches wide.

GLOSSARY (cont.)

Bearing Partition: An interior divider or wall that supports the structure above it.

Bearing Wall: A wall having weight-bearing properties associated with holding up a building's roof or second floor.

Bell: That portion of a pipe which is sufficiently enlarged to receive the end of another pipe of the same diameter.

Bell-and-Spigot Joint: Commonly used joint in cast-iron soil pipe end.

Benching: Making steplike cuts into a slope used for erosion control.

Benchmark: Point of known elevation from which surveyors can establish grades.

Berm: A raised earth embankment; the shoulder of a paved road; the area between the curb and the gutter and a sidewalk.

Bevel: A tool that can be adjusted to any angle; it helps make cuts at the number of degrees that is desired.

Bevel Siding: A siding material which is tapered from a thick edge to a thinner edge.

Bibb: A faucet used to connect a hose.

Bi-Fold: A double-leaf door used primarily for closet doors.

Bird Mouth: A notch cut into a roof rafter so that it can rest smoothly on the top plate.

Bitumen: The term used to identify asphalt and coal tar.

Black Pipe: Nongalvanized steel pipe.

Blank Flange: A soil plate flange used to seal off the flow in a pipe.

Blind Flange: A flange used to seal off the end of a pipe.

Blistering: The condition that paint presents when air or moisture is trapped underneath.

Blocking: A piece of wood fastened between structural members to strengthen them.

Board Foot (B.F.): A unit of lumber measure equaling 144 cubic inches; the base unit is 1 inch thick and 12 inches square or $1 \times 12 \times 12 = 144$ cubic inches.

Boiler Blow-Off: An outlet on a boiler to permit discharge of sediment.

Boiler Blow-Off Tank: A vessel designed to receive the discharge from a boiler blow-off outlet.

Bolster: A bent-wire device used in holding reinforcing bars in place during the pouring of concrete.

GLOSSARY *(cont.)*

Bond: In masonry, the interlocking system of brick or block to be installed.

Bond Beam: A steel-reinforced concrete masonry beam running horizontally around a masonry wall to provide added strength. Vertical bond beams are formed by inserting reinforcing bars in a cell after the wall is laid and filling with grout.

Bonding: A very-low-impedance path accomplished by permanently joining non-current-carrying metal parts is made to provide electrical continuity and to conduct current safely.

Bonding: The process of joining two surfaces together.

Bonding Conductor: The conductor that connects the non-current-carrying parts to the approved system ground conductor.

Bonding Jumper: A conductor used to connect the metal parts of an electrical system.

Bond Wire: Bare grounding wire that runs inside an armored cable.

Bonnet: Connects the valve actuator to the valve body.

Borrow Site: An area from which earth is taken for hauling to a jobsite which is short of earth.

Bottom or Heel Cut: The cutout of a rafter end which rests against the plate.

Bow: A term used to indicate an upward warp along the length of a piece of lumber.

Bow Window: A window unit that projects from an exterior wall.

Brace: An inclined piece of lumber applied to a wall or to roof rafters to add strength.

Branch: Any part of the piping system other than a main, riser, or stack.

Branch Circuit: Conductors between the last overcurrent device and the outlets.

Branch Circuit, Multiwire: A branch circuit having two or more ungrounded circuit conductors, each having a voltage difference between them, and a grounded circuit conductor (neutral) having an equal voltage difference between it and each ungrounded conductor.

Branch Interval: A length of soil or waste stack corresponding to a story height within which the horizontal branches are connected to the stack.

GLOSSARY *(cont.)*

Branch Tee: A tee having one side branch.

Branch Vent: A vent connecting one or more individual vents with a vent stack or stack vent.

Brazed: Joined by hard solder.

Brazing Ends: The ends of a valve or fitting which are prepared for silver brazing.

Break: The number of separate places on a contact that open or close a circuit.

Breakout Box: A device that allows access to individual points on a physical interface connector for testing.

Brick Veneer: A brick wall of single brick, usually covering a frame structure.

Bridging: Used to keep joists from twisting or bending.

British Thermal Unit (BTU): The amount of heat necessary to raise the temperature of one pound of water 1°F.

Bronze Trim or Bronze Mounted: Indicates that certain internal parts of the valves are made of copper alloy.

Bubble Tight: The condition of a valve seat that prohibits the leakage of visible bubbles.

Builder's Level: A tripod-mounted device that uses optical sighting to make sure that a straight line is sighted and reference point is level.

Building Codes: Rules and regulations instituted by a local housing authority or governing body.

Building Drain: The lowest piping of the drainage system which receives the discharge from waste, etc., inside the building and conveys it to the sewer.

Building Paper: Also called tar paper, roofing paper, etc.

Building Permits: Must be obtained for construction and allows for inspections of the work and for placing project on the tax roles.

Building Sewer: That part of the horizontal piping of a drainage system which extends from the end of the building drain and conveys it to any sewer.

Building Sewer, Combined: Conveys both sewage and storm water.

Building Sewer, Sanitary: Conveys sewage only.

Building Sewer, Storm: Conveys storm water only.

Building Subdrain: That portion of a drainage system which cannot drain by gravity in a building sewer.

Building Trap: A fitting or assembly of fittings installed in a building drain to prevent circulation of air between the drainage of the building and the building sewer.

Bull Head Tee: A branch of the tee is larger than the run.

Burst Pressure: The pressure which can be slowly applied to the valve at room temperature for 30 seconds without causing rupture.

Bus: A group of conductors that serve as a common connection for circuits.

Bus Bar: A heavy copper or aluminum bar used to carry currents in switchboards.

Bushing: A pipe fitting for connecting a pipe with a female or larger-size fitting.

Busway: A metal-enclosed distribution of bus bars.

Butt: To meet edge to edge.

Butterfly Valve: A device which operates at right angles to the flow.

Butt-Weld Joint: A welded pipe joint made with the ends of two pipes butting each other.

Butt-Weld Pipe: Pipe welded along a seam butted edge to edge and not scarfed or lapped.

By-Pass: An auxiliary loop in a pipeline intended for diverting flow around a device.

By-Pass Valve: A valve used to divert the flow past the part of the system through which it normally passes.

C

Cable: One or more insulated or noninsulated wires used to conduct electrical current.

Calcium Chloride: A concrete admixture used for accelerating the cure time.

California Bearing Ratio (CBR): A system used for determining the bearing capacity of a foundation.

Calorie (Cal): The amount of heat required to raise one gallon of water 1°C.

Camber: A slight vertical curve (arch) formed in a beam or girder to counteract deflection due to loading.

Cantilever: A projecting structural member or slab supported at one end only.

Cant Strip: A wooden strip used to raise the first course of shingles in plane; an angular board placed at the junction of the roof deck and wall to relieve the sharp angle when the roofing material is installed.

Capacitance (C): The ability of a circuit or component to store an electrical charge, measured in farads (F).

Capacitive Circuit: A circuit in which current leads voltage.

Capacitive Reactance (Xc): The opposition to current flow by a capacitor, in ohms (Ω).

Capacitor: A device that stores electrical energy by an electrostatic field.

Capacity: The maximum or minimum flows possible under given conditions of media, temperature, pressure, velocity, etc.

Capillary: The action by which the surface of a liquid, where it is in contact with a solid, is elevated or depressed.

Carbon Steel Pipe: Owes its properties mostly to the carbon it contains.

Carriage: A notched stair frame.

Casement: A type of window hinged to swing outward.

Casing: The trim that goes on around the edge of a door or window opening.

Catch Basin: A complete drain box where water drains into a pit, then through a pipe connected to the box.

Catch Point: Another name for hinge point or top of shoulder.

Cathode: The negative electrode in a battery.

Cathodic Protection: The control of the electrolytic corrosion of an underground or underwater metallic structure by the application of an electric current.

Caulk: Any type of material used to seal walls, windows, and doors to weatherize.

Cavitation: A localized gaseous condition that is found within a liquid stream.

Cement: A material that is the basis for a concrete mix.

Cementitious: Able to harden like cement.

Cement Joint: The union of two fittings by insertion of material.

Cement Plaster: A mixture of gypsum, cement, hydrated lime, sand, and water, used primarily for exterior wall finish.

Center Line: The point on stakes or drawings which indicates the halfway point between two sides.

Cesspool: A lined excavation in the ground which receives the discharge of a drainage system so as to retain the organic matter and solids but permit the liquids to seep through the bottom and sides.

Chainwheel-Operated Valve: A device which opens and closes valve seats.

Chair: Small device used to support horizontal rebar.

Chamfer: A beveled outside corner or edge on a beam or column.

Chase: A recess in a wall in which pipes can be run.

Check Valve: A device designed to allow a fluid to pass through in one direction only.

Chemical Waste System: Piping which conveys corrosive or harmful wastes to the drainage system.

Choke Coil: An inductor used to limit the flow of AC.

Chord: Top or bottom member of a truss.

Circuit: A complete path through which electricity flows.

Circuit Breaker: A device used to open and close a circuit.

Circuit Vent: A branch vent that serves two or more traps and extends from in front of the last fixture connection of a horizontal branch to the vent stack.

Circular Mil (cm): A measurement of the cross-sectional area of a conductor.

Clamp Gate Valve: A gate valve whose body and bonnet are held together by a U-bolt.

Cleanout: A plug joined to an opening in a pipe which can be removed for the purpose of cleaning.

Clear and Grub: To remove all vegetation, trees, concrete, or anything that will interfere with construction.

Clear Water Waste: Cooling water and condensate drainage from HVAC/R, cooled condensate from steam heating, cooled boiler blowdown water and waste water drainage in which impurities have been reduced below a minimum concentration considered harmful.

Cleat: Any strip of material attached to the surface of another material to support a third material.

Clerestory: A windowed area between roof planes or rising above lower story, to admit light and/or ventilation.

Closed Circuit: A continuous path for electrical flow.

Close Nipple: A nipple with a length twice the length of a standard pipe thread.

Cock: A form of valve having a hole in a tapered plug which is rotated to provide passageway for fluid.

Coefficient of Expansion: The increase in unit length, area, or volume for a one degree rise in temperature.

Coil: A winding of insulated conductors arranged to produce magnetic flux.

Cold Joint: Construction joint in concrete occurring at a place where the continuous pouring has been interrupted.

Collar Tie: Horizontal framing member tying the raftering together above the plate line.

Column: A vertical structural member.

Combined Waste and Vent System: A specially designed system of waste piping, embodying the horizontal wet venting of one or more floor sinks or floor drains by means of a common waste and vent pipe, adequately sized to provide free movement of air above the flow line of the drain.

Common Noise: Noise produced between the ground and the hot or neutral line.

Common Rafter: A structural member that extends without interruption from the ridge to the plate line in a sloped roof structure.

Common Vent: A vent which connects at the junction of two fixture drains and serves as a vent for both fixtures.

Compactor: A machine for compacting soil.

Companion Flange: A pipe flange to connect with another flange or with a flanged valve or fitting.

Compression Joint: A multi-piece joint with cup-shaped threaded nuts which, when tightened, compress tapered sleeves so that they form a tight joint.

Compressor: A mechanical device for increasing the pressure of air or gas.

Concrete: A mixture of sand, gravel, and cement in water.

Condensate: Water which has liquified from steam.

Condensation: The process by which moisture in the air becomes water or ice on a surface (such as a window) whose temperature is colder than the air's temperature.

Conductance: A measure of the ability of a component to conduct electricity in mohs.

Conductor: A substance which offers little resistance to the flow of electrical currents. The piping from the roof to the building storm drain or combined sewer, located inside of the building.

Conduit: Metal, fiber pipe or raceway used to carry electrical conductors.

Conduit Body: The part of a conduit system, at the junction of two or more sections of the system, that allows access through a removable cover.

Confluent Vent: A vent serving more than one fixture vent or stack vent.

Construction Joint: Separation between two placements of concrete; a means for keying two sections together.

Contactor: A control device that uses a small current to energize or de-energize.

Contacts: The conducting part of a switch that operates with another conducting part to make or break a circuit.

Continuous Load: A load whose maximum current continues for three hours.

Continuous Vent: A vent that is a continuation of the drain to which it connects.

Continuous Waste: A continuous drain from two or three fixtures connected to a single trap.

Contour Line: Solid or dashed lines showing the elevation of the earth.

Control: A device used to regulate the function of a component or system.

Convection: Transfer of heat through the movement of a liquid or gas.

Convector: A heat transfer device (radiator) used in a hydronic (hot water) system.

Coping: The top course or cap on a masonry wall protecting the masonry below.

Corbel: A stone, masonry, or wood bracket projecting out from a wall.

Corner Beads: Metal strips that prevent damage to drywall corners.

Cornice: That part of the roof extending horizontally out from the wall.

Corporation Cock: A stopcock screwed into the street water main to supply a house service connection.

Coulomb: Current of one ampere per second.

Coupling: A pipe fitting with female threads used to connect two pipes.

Course: A horizontal layer of masonry units.

Crawl Space: The area under a floor that is only excavated to allow one to crawl under it.

Cripple Jack: A jack rafter with a cut that fits in between a hip and a valley rafter.

Cripple Rafter: A cripple rafter is not as long as the regular rafter.

Cripple Stud: A short stud that fills out the position where the stud would have been located if a window, door, or some other opening had not been there.

Critical Level: The point on a backflow-prevention device or vacuum breaker marked C/L which determines the minimum elevation above the flood-level rim of the fixture served at which the device may be installed; when a backflow-prevention device is not marked C/L, the bottom of the vacuum breaker or combination valve constitutes the critical level.

Cross: A pipe fitting with four branches in pairs, each pair on one axis.

Cross Brace: Wood or metal diagonal bracing used to aid in structural support between joists and beams.

Cross Connection: Any physical connection between two otherwise separated piping systems, one of which contains potable water and the other of unknown safety, whereby flow may occur from one system to the other.

Crossover: A pipe fitting with a double offset or shaped like the letter "U" with ends turned out; used to pass the flow of one pipe past another.

Crosstalk: The unwanted energy transferred from one circuit or wire to another.

Cross Valve: A valve fitted on a transverse pipe so as to open communication between two parallel pipes.

Crown: The top of a trap.

Crown Vent: A vent pipe connected at the uppermost point in the crown of a trap.

Crows Foot: A lath set by the grade setter with markings to indicate the final grade at a certain point.

Cup: To warp across the grain.

Cup Weld: A pipe weld in which one pipe is expanded on the end to allow the entrance of the end of the other pipe.

Curb Box: A device at the curb that contains a valve used to shut off a supply line.

Current: The flow of electricity in a circuit, in AMPS.

Curtain Wall: Inside walls that do not carry loads.

Cutout Box: A surface-mounted electrical enclosure with a hinged door.

Cutting Plane Line: A heavy broken line with arrows, letters, and numbers at each end indicating the section view that is being identified.

Cycle: Measured in hertz (Hz), it is the flow of AC in one direction and then in the opposite direction in one time interval.

D

Dado: A rectangular groove cut into a board across the grain.

Daisy Chaining: The connection of multiple devices in a serial fashion.

Dampen: To check or reduce.

Dampproofing: A surfacing used to coat and protect concrete and masonry from moisture.

Data Rate: The number of bits of information in a transmission system, expressed in bits per second.

Datum Point: See Benchmark; identification of the elevation above mean sea level.

DC Compound Motor: The field is connected in both series and shunt with the armature.

DC Permanent Magnet Motor: Uses magnets, not a coil, for the field winding.

DC Series Motor: The field is connected in series with the armature.

DC Shunt Motor: The field is connected in parallel with the armature.

DC Voltage: Voltage that flows in one direction only.

Dead End: A branch leading from a soil, waste, or vent pipe, a building drain, or a building sewer which is terminated at a developed distance of 2 feet or more by means of a plug, etc.

Dead Load: The weight of a structure and all its fixed components.

Decibel: A standard logarithmic unit for the ratio of two powers, voltages, or currents. In fiber optics, the ratio is power.

Deck: The part of a roof that covers the rafters.

Deep Cycle: Battery type that can be discharged to a large fraction of capacity.

Deformed Bar: Steel reinforcement bar with ridges to prevent the bar from loosening during the concrete curing process.

Degauss: To remove residual permanent magnetism.

Delta Connection: A connection that has each coil connected end-to-end.

Department Having Jurisdiction: The administrative authority affected by any provision of a building code.

Depth of Discharge (DOD): The percent of the rated battery capacity that has been withdrawn.

Developed Length: The length along the center line of a pipe and fittings.

Device (Wiring Device): The part of an electrical system that is designed to carry, but not use, electrical energy.

Dewpoint: The temperature of a gas or liquid at which condensation or evaporation occurs.

Diagonal Brace: A wood or metal member placed diagonally over wood or metal framing to add rigidity at corners and at 25'0" of unbroken wall space.

Diaphragm: A flexible disk which actuates the valve stem.

Diaphragm Control Valve: A control valve having a spring diaphragm actuator.

Dielectric Fitting: A fitting having insulating parts or material that prohibits flow of electrical current.

Differential: The variance between two target values, one is high, the other low.

Diffuser: A grille or register over the air duct opening into a room which controls and directs the flow of air.

Digestion: The portion of the sewage treatment process where biochemical decomposition of organic matter takes place, resulting in the formation of simple organic and mineral matter.

Dimension Line: A line on a drawing with a measurement indicating length.

Direct Current (DC): Electrical current which flows in one direction only.

Disk: That part of a valve which actually closes off the flow.

Displacement: The volume or weight of a fluid displaced by a floating body.

Diverter: A piece, usually metal, used to direct moisture to a desired path or location.

Domestic Sewage: Liquid and water-borne wastes that are free from industrial wastes and permit satisfactory disposal without special treatment into a public or private sewer.

Dormer: A projection built out from a sloping roof, including one or more vertical windows.

Dosing Tank: A watertight tank in a septic system placed between the tank and the distribution box equipped with a pump or automatic siphon designed to discharge sewage to a disposal field.

Double Break Contacts: Contacts that break the current in two separate places.

Double Disk: A two-piece disk used in a gate valve.

Double Offset: Two changes of direction installed in succession or series in continuous pipe.

Double Plate: Usually refers to the practice of using two pieces of dimensional lumber for support over the top section or wall section.

Double Ported Valve: A valve having two parts to overcome line-pressure imbalance.

Double-Sweep Tee: A tee made with long-radius curves between body and branch.

Double Trimmer: Double joists used in the sides of openings placed without regard to regular joist spacings for stairs or chimneys.

Double Wedge: A device used in gate valves, similar to a double disk, in which the split wedges seal independently.

Dowel: Straight metal bars used to connect or position two sections of concrete or masonry.

Down: Refers to piping running through the floor to a lower level.

Downspout: The rainleader from the roof to a building storm drain.

Downstream: Refers to a location in the direction of flow after passing a reference point.

Drain: Any pipe which carries waste water or water-borne wastes in a drainage system.

Drainage Fitting: A type of fitting used for draining fluid from pipes and making a smooth and continuous interior surface.

Drainage System: The drainage piping within public or private premises which conveys sewage, rain water, or other liquid wastes to an approved point of disposal, but not including the mains of a public sewer system.

Drain Field: An area of a piping system arranged in troughs for disposing liquid waste.

Drain Tile: Usually 4" plastic pipe with small holes that allow water to drain into it; laid along the foundation footing to drain the seepage into a sump or storm sewer.

Droop: The amount by which the controlled variable pressure, temperature, liquid level, or differential pressure deviates from a set value.

Drop: Refers to piping running to a lower elevation within the same floor level.

Drop Elbow: A small elbow having wings cast on each side to secure to a ceiling, wall, etc.

Drop Siding: Has a special groove cut into it that lets each board fit into the next board.

Drop Tee: A tee having wings as in a drop elbow.

Dross: Solid scum that forms on the surface of a metal when molten or melting as a result of oxidation or dirt.

Dry Wall: A type of wall covering (gypsum board) used in place of plaster.

Dry-Pipe Valve: A valve used with a dry-pipe sprinkler system in which water is on one side and air is on the other.

Dry-Weather Flow: Sewage collected during the summer which contains little or no ground water by infiltration and no storm water.

Duct: A round or rectangular pipe, usually metal, used for transferring conditioned air in a heating and cooling system.

Ductwork: A system of pipes used to distribute air to all parts of a structure.

Durham System: A term used to describe soil or waste systems in which all piping is of threaded pipe, tubing, or other rigid construction.

DWV: Type of copper or plastic tubing used for drain, waste, or venting pipe.

E

Earthwork: Excavating and grading soil.

Easement: A portion of land on or off a property which is set aside for utilities.

Eave: The lowest edge on a gable roof.

Eaves: The overhang of a roof projecting over walls.

Eaves Trough: A gutter.

Eccentric Fittings: Fittings whose openings are offset and allow liquid to flow freely.

Effective Openings: The minimum cross-sectional area at the point of discharge.

Efficiency (EFF): The ratio of output power to input power, expressed in percent.

Effluent: Sewage, treated or partially treated, flowing out of sewage treatment equipment.

Elastic Limit: The greatest stress which a material can withstand without a permanent deformation after release of the stress.

Elbow: A fitting that makes an angle between adjacent pipes.

Electricity: The movement of electrons through a conductor.

Electrolysis: The process of producing chemical changes by passage of an electric current through an electrolyte.

Electromagnet: Coil of wire that exhibits magnetic properties when current passes through it.

Elevation: An exterior or interior orthographic view of a structure, identifying the design and the materials to be used.

Elevation Numbers: The vertical distance above or below sea level.

Embankment: Area being filled with earth.

End Connection: The method of connecting the parts of a piping system.

Engineered Plumbing System: Designed by using engineering design criteria other than those given in plumbing codes.

Equalization: The process of restoring all cells in a battery to an equal state of charge.

Erosion: The gradual destruction of metal or other material by the abrasive action of liquids, gases, etc.

Evapotranspiration: Loss of water from the soil by evaporation and plants.

Excavation: The recess or pit formed by removing the earth in preparation for footings, etc.

Excitation: The power required to energize the

GLOSSARY (cont.)

magnetic field of motors, transformers, generators, etc.

Existing Work: A plumbing system which has been installed prior to the effective date of applicable code.

Expansion Joint: A joint whose primary purpose is to absorb longitudinal thermal expansion in a pipe line.

Expansion Joint: Formed in concrete or masonry units by a bituminous fiber strip to allow for expansion and contraction.

Expansion Loop: A large-radius bend in a pipe line to absorb longitudinal expansion.

Extra Heavy: Description of piping material, usually cast iron, indicating thicker than standard.

Extrusion: Metal which has been shaped by forcing it in the hot or cold state through dies of the desired shape.

F

Face: The exposed side of a framing or masonry unit.

Face Brick: A select brick fired to produce a desired color and effect for use in the face of a wall.

Face-to-Face Dimensions: The dimensions from the face

of the inlet port to the face of the outlet port of a valve, etc.

Farad (F): The unit of measurement of capacitance.

Fascia: A flat board covering the ends of rafters on the cornice or eaves.

Fault Current: Any current that travels an unwanted path.

Feathering: Raking new asphalt to join smoothly with the existing asphalt.

Feeder: Circuit conductors between the service and the final branch circuit OCPD.

Female Thread: Internal thread in pipe fittings, etc.

Ferrule: A precision tube that holds a fiber for alignment.

Fiber Optics: A technology that uses light as a digital information carrier.

Field: The stationary windings (magnets) of a DC motor.

Filament: A conductor that has a high enough resistance to cause heat.

Filter: A combination of circuit elements designed specifically to pass certain frequencies and resist all others.

Filter: Device through which fluid is passed to separate contaminates from it.

Filter Element or Media: A porous device which performs the process of filtration.

Finish: Any material used to complete an installation that provides an esthetic or finished appearance.

Firebrick: A special type of brick that is not damaged by fire; used to line the firebox.

Fire Hydrant Valve: A valve that drains at an underground level to prevent freezing.

Fire Pumps:

Can Pump: A vertical-shaft turbine-type pump used to raise water pressure.

Centrifugal Pump: The pressure is developed by centrifugal force.

End Suction Pump: A single suction pump having its suction nozzle on the opposite side of the casing from the stuffing box and having the face of the suction nozzle perpendicular to the shaft.

Excess Pressure Pump: Low-flow, high-head pump for sprinkler systems not being supplied from a fire pump.

Fire Pump: A pump with driver, controls, and accessories used for fire protection service; fire pumps are centrifugal or turbine type.

Horizontal Pump: The shaft is in a horizontal position.

Horizontal Split-Case Pump: A centrifugal pump characterized by a housing which is split parallel to the shaft.

In-Line Pump: A centrifugal pump whose drive unit is supported by the pump, having its suction and discharge flanges on approximately the same center line.

Pressure Maintenance (Jockey) Pump: A pump used to maintain pressure in a fire protection system without the operation of the fire pump.

Vertical-Shaft Turbine Pump: A centrifugal pump with one or more impellers discharging into one or more bowls and a vertical educator or column pipe used to connect the bowl(s) to the discharge head on which the pump driver is mounted.

Fire Stop/Draft Stop/Fire Blocking: A framing member used to reduce the ability of a fire to spread.

Firewall/Fire Separation Wall/Fire Division Wall: Any

wall that is used to prevent the spread of fire.

Fitting: The connector or closure for fluid lines.

Fitting, Compression: A fitting designed to join pipe or tubing by means of pressure.

Fitting, Flange: A fitting which utilizes a radially extending collar for sealing.

Fitting, Welded: A fitting attached by welding.

Fixture Branch: A pipe connecting several fixtures.

Fixture Carrier: A metal unit designed to support a plumbing fixture off the floor.

Fixture Carrier Fittings: Special fittings for wall-mounted fixture carriers.

Fixture Drain: The drain from the trap of a fixture to the junction of that drain with any other drain pipe.

Fixture Supply: A water-supply pipe connecting the fixture with the fixture branch or directly to a water main.

Fixture Unit: A measure of probable discharge into a drainage system by various types of plumbing fixtures.

Fixture Unit Flow: A measure of the probable hydraulic demand on a water supply by various types of plumbing fixtures.

Flange: A ring-shaped plate on the end of a pipe at right angles to the end of the pipe and provided with holes for bolts to allow fastening.

Flange Bonnet: A valve bonnet having a flange which bolts to a matching flange on the valve body.

Flange Ends: A valve or fitting having flanges for joining to other piping, etc.

Flange Faces: Pipe flanges which have the entire surface faced straight across, using a full face or ring gasket.

Flanges: The parallel faces of a structural beam joined by the web of the beam.

Flap Valve: A Nonreturn valve in the form of a hinged disk or flap.

Flashing: Metal or plastic strips or sheets used for moisture protection in conjunction with other construction materials.

Flashover: A disruptive electrical discharge around or over an insulator.

Flash Point: The temperature at which a fluid gives off sufficient flammable vapor to ignite.

Flat: In roofing, any roof structure up to a 3:12 slope.

Float Valve: A valve which is operated by means of a bulb or ball floating on the surface of a liquid within a tank.

Flooded: A condition when the liquid rises to the flood-level rim of a fixture.

Flood-Level Rim: The top edge of a plumbing receptacle, from which water overflows.

Flow Pressure: The pressure in a water supply pipe near the water outlet while the faucet or outlet is fully open.

Flue: An enclosed passage, normally vertical, for removal of gaseous products of combustion to the outer air.

Fluorescence: The emission of light by a substance when exposed to radiation.

Flush: To be even with.

Flush Door: A smooth-surface door without panels or molding.

Flushing-Type Floor Drain: A floor drain which is equipped with an integral water supply, enabling flushing.

Flushometer Valve: A device which discharges a predetermined quantity of water to fixtures for flushing purposes.

Flux: An electrical field energy distributed in space and represented diagrammatically by means of flux lines denoting magnetic or electrical forces.

Fly Ash: Fine, powdery coal residue used with a hydraulic (water-resistant) concrete mix.

Footcandle (fc): The amount of light produced by a lamp measured in lumens divided by the area that is illuminated.

Footing: The part of a foundation wall or column resting on the bearing soil, rock, or piling which transmits the superimposed load to the bearing material.

Foot Valve: A check valve installed at the base of a pump suction pipe to maintain pump prime.

Form: A temporary construction member used to hold permanent materials in place.

Foundation: The base on which a house or building rests.

Four-Wire Circuits: Telephone circuits which use two separate one-way transmission paths of two wires each.

Framing: The wood or metal structure of a building which gives it shape and strength.

French Drain: A drain consisting of an underground passage made by filling a trench with loose stones.

Frequency: The number of times per second a signal regenerates itself at a peak amplitude.

Fresh-Air Inlet: A vent line connected with the building drain just inside the house trap and extending to the outer air providing fresh air to the lowest point of a plumbing system.

Frostline: The depth to which ground freezes.

Frostproof Closet: A hopper that has no water in the bowl and has the trap and the control valve for its water supply installed below the frost line.

Full-Load Current (FLC): The current required by a motor to produce the full-load torque at the motor's rated speed.

Full-Load Torque (FLT): The torque required to produce the rated power of the motor at full speed.

Furring Strips: Strips of wood attached to concrete or stone that form a nail base for wood or paneling.

Fuse: A protective device, also called an OCPD.

Fusion Weld: Joining metals by fusion, using oxyacetylene or electric arc.

G

Gable: The simplest kind of roof; two large surfaces come together at a common edge, forming an inverted V.

Gain: A ratio of the amplitude of the output signal to the input signal.

Galvanize: A coating of zinc used primarily on sheet metal or pipe.

Gambrel Roof: A barn-shaped roof.

Gate Valve: A valve employing a gate, often wedge-shaped, allowing fluid to flow when the gate is lifted from the seat.

Gauge: The thickness of metal, glass or wire.

Ghost Voltage: A voltage that appears on a motor that is not connected.

Girder: A support for joists at one end; usually placed halfway between the outside walls and runs the length of the building.

Glaze: To install glass.

Globe Valve: Globe-shaped body with a manually raised or lowered disc.

Glu-lam (GLB): Beam made from milled 2x lumber bonded together.

Grade: An existing or finished elevation in earthwork; a sloped portion of a roadway; sizing of gravel and sand; the structural classification of lumber.

Grade: The slope or fall of a line of pipe with reference to a horizontal plane; in drainage it is expressed as the fall in a fraction of an inch.

Grade Beam: A low foundation wall or a beam, usually at ground level, which provides support for the walls of a building.

Grade Break: A change in slope from one incline ratio to another.

Grade Lath: A piece of lath that the surveyor marked to indicate the correct grade to the operators.

Grade Pins: Steel rods driven into the ground at each surveyor's hub.

Grader: A power excavating machine with a central blade that can be angled to cast soil on either side.

Gravel Stop: The edge metal used at the eaves of a built-up roof to hold the gravel.

Green: Uncured or set concrete or masonry; freshly cut lumber.

Grid: An electrical utility distribution network.

Grid System: A system of metal strips that supports a drop ceiling.

Ground: An electrical connect between equipment and the earth.

Ground Fault: Current from a hot line is flowing to the ground.

Ground-Fault Circuit Interrupter (GFCI): An electrical device which protects personnel by detecting hazardous ground faults and quickly disconnects power from the circuit.

Grounding: The connection of all exposed non-current-carrying metal parts to earth.

Ground Joint: The parts are precisely finished and then ground in so that the seal is tight.

Grout: A cementitious mixture of high water content made from Portland cement, lime, and aggregate, used to secure anchor bolts and vertical reinforcing rods in masonry walls.

GLOSSARY *(cont.)*

Guinea: A survey marker driven to grade; it may be colored with paint.

Gusset: A triangular or rectangular piece of wood or metal that is usually fastened to the joint of a truss to strengthen it.

Gutter: A metal trough set below the eaves to catch and conduct water to a downspout.

Guy: A wire having one end secured and the other fastened to a pole or structure under tension.

Gypsum: A chalk used to make wallboard; made into a paste, inserted between two layers of paper and allowed to dry.

H

Habitable Space: In residential construction, the interior areas of a residence used for eating, sleeping, living, and cooking; excludes bathrooms, storage rooms, utility rooms, and garages.

Hanger: Metal fabrication made for placing and supporting joists and rafters.

Hardware: Any component used to hang, support, or position another component.

Hardwood: Wood from a tree that sheds it leaves.

Haunch: Portion of a beam that increases in depth toward the support.

Header: A framing member used to hide the ends of joists along the perimeter of a structure; also known as a rim joist; the horizontal structural framing member installed over wall openings to aid in the support of the structure above.

Header: A large pipe or drum into which each of a group of boilers is connected; also used for a large pipe from which a number of smaller ones are connected in line from the side of the large pipe.

Header Course: In masonry, a horizontal row of brick laid perpendicular to the wall face; used to tie a double-wythe brick wall together.

Head Joint: The end face of a brick or concrete masonry unit to which the mortar is applied.

Heater: A device that is placed in a motor starter to measure the amount of current in the power line.

Heating Element: A conductor (wire) that offers enough resistance to produce heat when connected to power.

GLOSSARY *(cont.)*

Henry (H): The unit of measure of inductance.

Hertz (Hz): One Hertz is equal to one cycle of the AC sine wave per second.

Hidden Line: A dashed line identifying portions of construction that are a part of the drawing but cannot be seen; e.g., footings on foundation plans.

HID Lamp: High-intensity discharge lamp.

Hip Rafters: A member that extends diagonally from the corner of the plate to the ridge.

Hip Roof: A structural sloped roof design with sloped perimeters from ridge to plate line.

Hollow-Core Door: A lightweight flush door with an interior core of glued strips forming a honeycomb and two exterior smooth panels.

Honeycomb: Voids or open spaces left in concrete due to a loss or a shortage of mortar.

Horizontal Branch: A drain pipe extending laterally from a soil, waste stack, or drain.

Horsepower (HP): A unit of power equal to 746 watts that describes the output of electric motors.

Hose Bibb: A faucet used to connect a hose.

Hub: A device which connects to several other devices, usually in a star topology.

Hub and Spigot: Piping made with an enlarged diameter or hub at one end and plain or spigot at the other end; the joint is made tight by oakum, lead, or gasket.

Hubless: Soil piping with plain ends; the joint is made tight with a clamp and gasket.

HVAC: Heating, Ventilating and Air Conditioning.

Hybrid: An electronic circuit that uses different cable types to complete the circuit.

Hydraulic Cement: A cement used in a concrete mix capable of curing under water.

I

Impedance (Z): The total opposition offered to the flow of AC from resistance and reactance measured in ohms (Ω).

Increaser: A pipe coupling used between pipes of different sizes.

Indirect Waste Pipe: A pipe that does not connect directly with the drainage system but discharges into a plumbing fixture or receptacle that is connected directly to the drainage system.

Individual Vent: A pipe installed to vent a fixture trap that connects with the vent system above the fixture served or terminates in air.

Induced Siphonage: Loss of liquid from a fixture trap due to pressure differential between the inlet and outlet of the trap.

Inductance (L): The property of a circuit that determines how much voltage will be induced into it by a change in current of another circuit; measured in henrys (H).

Inductive Circuit: A circuit in which current lags voltage.

Inductive Reactance (X_L): The opposition to the flow of AC in a circuit due to inductance, measured in ohms (Ω).

Industrial Waste: All liquid or water-borne waste from industrial or commercial processes.

In-Phase: The state when voltage and current reach maximum amplitude and zero level at the same time.

Insanitary: A condition which is contrary to sanitary principles or is unhealthy.

Insulating Glass: A window or door glass consisting of two sheets of glass separated by a sealed air space.

Insulation: Any material capable of resisting thermal, sound, or electrical travel.

Insulation Resistance: The R factor in insulation calculations.

Insulator: Material in which current cannot flow easily.

Integrally Cast: Element (such as concrete joist and top slab) cast in one piece. See Monolithic.

Interceptor: A device that separates and retains hazardous or undesirable matter from normal wastes and permits normal liquid wastes to discharge into the disposal terminal by gravity.

Interface: The point at which two systems connect.

Invert: The lowest point on the interior of a horizontal pipe.

Isolated Grounded Receptacle: Minimizes electrical noise by providing a separate grounding path.

Isolation Transformer: A one-to-one transformer used to isolate equipment at the secondary from earth ground.

Isometric Projection: A pictorial drawing positioned so that its principal axis make equal angles with the plane of projection.

J

Jack: A receptacle (female) used with a plug (male) to make a connection.

Jacket: The protective and insulating outer housing on a cable. Also called a sheath.

Jack Rafter: A part of the roof structure raftering that does not extend the full length from the ridge beam to the top plate.

Jamb: The part that surrounds a door or window frame.

Joint Compound: Material used with a paper or fiber tape for sealing indentations and breaks in drywall construction.

Joist: A structural horizontal framing member used for floor and ceiling support.

Joist Hangers: Metal brackets that hold up the joist.

Joule: A unit of electrical energy also called a watt-second.

Jumper: Patch cable or wire used to establish a circuit, often temporarily, for testing.

Junction Box: A box, usually metal, that encloses cable connections.

K

Kalamein Door: A metal-covered, fireproof door.

Key: A depression made in a footing so that the foundation or wall can be poured into the footing, preventing the wall or foundation from moving during changes in temperature or settling.

Kicker Blocks: Cement poured behind each bend or angle of water pipe for support.

Kiln-Dried Lumber: Lumber that is seasoned under controlled conditions.

Kilo: The metric system prefix meaning one thousand.

King Stud: A full-length stud from the bottom plate to the top plate, supporting both sides of a wall opening.

Knee Wall: Vertical framing members supporting and shortening the span of the roof rafters.

kWH: Kilowatt hour. The basic unit of electrical energy for utilities, equal to one thousand watts of power supplied for one hour.

GLOSSARY *(cont.)*

L

Labeled: Equipment or materials bearing a label of a listing agency.

Lally Column: A vertical steel pipe, usually filled with concrete, used to support beams and girders.

Laminated Plastic: Layers of cloth or other fiber impregnated with plastic.

Lamination: A method of constructing by placing layer upon layer of material and bonding with an adhesive.

Lamp: A light source. Reference is to a light bulb.

Landing: A platform in a flight of stairs to change the direction or break a run.

Lapped Joint: A pipe joint made by using loose flanges on lengths of pipe whose ends are lapped over to produce a bearing surface for a gasket or metal-to-metal joint.

Lap Weld Pipe: Made by welding along a scarfed longitudinal seam in which one part is overlapped by the other.

Lateral: Underground electrical service.

Lateral Sewer: A sewer which does not receive sewage from any other common sewer except house connections.

Lath: Backup support for plaster; may be of wood, metal, or gypsum board.

Lavatory: Bathroom; vanity basin.

Lay-In Ceiling: A suspended ceiling system.

Leaching Well: A pit or receptacle having porous walls which permit the contents to seep into the ground; also called a dry well.

Leach Line: A perforated pipe used as a part of a septic system to allow liquid overflow to dissipate into the soil.

Leader: The water conductor from the roof to the building storm drain.

Lead Joint: A joint made by pouring molten lead into the space between a bell and spigot and making the lead tight by caulking.

Leakage Current: Current that flows through insulation.

Ledger: Structural framing member used to support ceiling and roof joists at the perimeter walls.

Leg (Circuit): One of the conductors in a supply circuit in which the maximum voltage is maintained.

GLOSSARY *(cont.)*

Level-Transit: An optical device that is a combination of a level and a means for checking vertical and horizontal angles.

Lift: Any layer of material or soil placed upon another.

Lift Slab: Concrete floor construction in which slabs are cast directly on one another secured by column brackets or collars.

Lintel: Support for a masonry opening, usually steel angles or special forms.

Lip Union: A union characterized by a lip that prevents a gasket from being squeezed into the pipe.

Liquid Waste: The discharge of a plumbing system which does not receive fecal matter.

Listed: Equipment or materials that comply with approved standards or has been tested and found suitable for use in a specified manner.

Live Load: Any movable equipment or personal weight to which a structure is subjected.

Load: The amount of electric power used by any electrical unit or appliance at any given moment.

Load: The weight of a building.

Load Conditions: The conditions under which a roof must perform.

Load Factor: The percentage of the total connected fixture unit flow which is likely to occur at any point in a drainage system.

Location, Damp: Partially protected locations, such as under canopies, roofed open porches, basements, barns, etc.

Location, Wet: Locations underground, in concrete slabs, where saturation occurs, or outdoors.

Locked Rotor: Condition when a motor is loaded so heavily that the shaft cannot turn.

Locked Rotor Current (LRC): The steady-state current with the rotor locked and the voltage applied.

Locked Rotor Torque (LRT): The torque a motor produces when the rotor is stationary and full power is applied.

Lockset: The doorknob and associated locking parts inserted in a door.

Longitudinal: The long dimension of an object.

Lookout: The structural member running from the outside wall to the ends of rafters to carry the plancier or soffit.

Louver: A ventilated opening in the attic, usually at a gabled end, made of inclined horizontal slats, to permit air to pass but to exclude moisture.

Lumen (lm): The unit used to measure the total amount of light produced by a source.

M

Magnetic Field: The invisible field produced by a current-carrying conductor, coil, etc. which develops a north and south polarity.

Magnetic Flux: The invisible lines of force that make up a magnetic field.

Main: The principal artery of a system of continuous piping, to which branches may be connected.

Main Vent: A vent header to which vent stacks are connected.

Malleable: Capable of being shaped by hammer or rolling pressure.

Malleable Iron: Cast iron that is heat-treated.

Manifold: A fitting with a number of branches in line connecting to smaller pipes.

Masonry: Manufactured materials of clay (brick), concrete (CMU), and stone.

Mastic: An adhesive used to hold tiles in place; also refers to adhesives used to glue many types of materials.

Mat: Asphalt as it comes out of a spreader box or paving machine in a smooth, flat form.

Maximum Density and Optimum Moisture: The highest point on the moisture–density curve; considered the best compaction of the soil.

Medium Pressure: Means that valves and fittings are suitable for a working pressure of 125 to 175 psi.

MEE Pipe: Pipe that has been milled on each end and left rough in the center; MEE stands for "milled each end."

Membrane Roofing: Built-up roofing.

Mesh: Common term for welded-wire fabric, plaster lath.

Mil: 0.001 inch.

Mill Length: Also known as random length; run-of mill pipe is 16 to 20 feet in length; some pipe is made in double lengths of 30 to 35 feet.

Minute: 1/60th of a degree.

MOA Pipe: Pipe that has been milled end to end; MOA stands for "milled over all".

Modular Measurement: The design of a structure to use standard-size building materials. In the customary system of measurement the module is 4 inches. In the metric system, the recommended module is 100 millimeters.

Moisture Barrier: A material used for the purpose of resisting exterior moisture penetration.

Moisture-Density Curve: A graph showing at what point of added moisture the maximum density occurs.

Moldings: Trim mounted around windows, floors, etc.

Monolithic Concrete: Concrete placed as a single unit including footings.

Mortar: A concrete mix used for bonding masonry.

Motor: A machine that develops torque on a shaft to produce work.

Motor Efficiency: The effectiveness of a motor to convert electrical energy into mechanical energy.

Motor Starter: An electrically operated switch (contactor) that includes overload protection.

Motor Torque: The force that produces rotation in a shaft.

Multiplex: To combine multiple input signals into one for transmission over a single high-speed channel.

N

Natural Grade: Existing or original grade elevation.

Neat Cement: A pure cement mixture, with no sand or other material added.

NEC: National Electrical Code, which contains safety rules for installations.

Needle Valve: A valve with a long tapering point in place of an ordinary valve disk.

Nipple: A tubular pipe fitting normally threaded on both ends and less than 12 inches in length.

No-Load Current: The current demand of a transformer primary when no current demand is made on the secondary.

Nominal Size: A general classification term used to designate size of commercial products, such as a 2 × 4. This is not an actual size.

Nominal Size: Original cut size of a piece of lumber prior to milling and drying; size of a masonry unit, including mortar bed and head joint.

Nonbearing: Not supporting any structural load.

Normally Closed Contacts: Contacts that are closed before being energized.

Normally Open Contacts: Contacts that are open before being energized.

Nuclear Test: A test to determine soil compaction by sending nuclear impulses into the compacted soil and measuring the returned impulses reflected from the compacted particles.

O

O.D. Pipe: Pipe that measures over 14 inches normal pipe size, where the nominal size is the outside diameter and not the inside diameter.

Offset: A combination of pipe and/or fittings which join two nearly parallel sections of a pipe line.

Ohm: The unit of measurement of electrical resistance.

Ohm's Law: A law which describes the mathematical relationship among voltage, current, and resistance.

On Center (O/C): The distance between the centers of two adjacent components.

Open Circuit: A condition that provides no path for electric current to flow in a circuit.

Open-Circuit Voltage: The maximum voltage produced without a load applied.

Open-Web Joist: Roof joist made of wood or steel construction with a top chord and bottom chord connected by diagonal braces. Some manufacturers make a joist with chords of wood and a steel web and refer to it as a truss joist.

Orthographic Projection: The basis of architectural plan and elevation drawings.

Oscillation: Fluctuations in a circuit.

Outfall Sewers: Sewers receiving sewage from a collection system and carrying it to the point of final discharge or treatment.

Outlet: Where the current is taken to supply equipment.

Overload Protection: A device that prevents overloading a circuit or motor.

Oxidized Sewage: Sewage in which the organic matter has been combined with oxygen, resulting in natural stability.

P

Package Air Conditioner or Boiler: All components are packaged in a single unit.

Pad: In earthwork or concrete foundation work, the base materials used on which to place the concrete footing and/or slab.

Panel Door: A door of solid frame strips with inset panels.

Parallel Circuit: More than one path through which current flows.

Parapet: An extension of an exterior wall above the line of the roof.

Parging: A thin moisture-protection coating of plaster or mortar over a masonry wall.

Partition: An interior wall separating two rooms or areas of building; usually nonbearing.

Penny (d): Unit of measure of nails used by carpenters.

Percolation: The seeping of a liquid downward through a filtering medium.

Perimeter: The outside edges of a plot of land or building; it represents the sum of all the individual sides.

Perimeter Insulation: Insulation placed around the outside edges of a slab.

Photovoltaic: Changing light into electricity.

Pier: A heavy column of masonry between two openings used to support other structural members.

Pilaster: A column projecting on the outside or inside of a masonry wall to add strength or decorative effect.

Pile: A steel or wooden pole driven into the ground sufficiently to support the weight of a wall and building.

Pillar: A pole or reinforced wall section used to support the floor and consequently the building.

Pitch: The amount of slope or grade given to horizontal piping and expressed in inches.

Pitch: The slant or slope from the ridge to the plate.

Plan View: A bird's-eye view of a construction layout cut at 5'0" above finish floor level.

Plancier: The board or panel forming the underside of the eave or cornice.

Plaster: A mixture of cement, water, and sand.

Plat: A drawing of a parcel of land indicating lot number, location, boundaries, and dimensions. Contains information as to easements and restrictions.

Plate: A roof member which has the rafters fastened to it at their lower ends.

Platform Framing: Also known as western framing; structural construction in which all studs are only one story high with joists over.

Plenum: A chamber in an A/C system which receives air under pressure before distribution to ducts.

Plug Valve: A valve with a short section of a cone or tapered plug.

Plumb: Perpendicular or vertical. Also: to make a structure vertical.

Plumbing: The practice, materials, and fixtures used in the installation, maintenance, and alteration of all piping, fixtures, appliances, etc, in connection with sanitary or storm drainage facilities; venting systems; public or private water supply systems within or adjacent to any building or structure.

Plumbing Appliance: A plumbing fixture which is intended to perform a special function; its operation may be dependent on one or more energized components; and

these fixtures may operate automatically through a time cycle, temperature range, pressure range, etc.

Plumbing Appurtenances: A manufactured device, prefabricated assembly, or job-assembled component which is added to a basic plumbing system.

Plumbing Fixtures: Installed receptacles, devices, or appliances which are supplied with water or which receive liquid or liquid-borne wastes and discharge such wastes into a drainage system.

Plumbing Inspector: Is authorized to inspect plumbing and drainage as defined in the code for the municipality.

Plumbing System: All potable water supply and distribution pipes, plumbing fixtures and traps, drainage and vent pipes, and all building (house) drains, devices, receptacles, and appurtenances within the property lines of the premises.

Pocket Door: A door which slides into a partition or wall.

Point of Beginning (POB): The point on a property from which all measurements and azimuths are established.

Polarity: The particular state of an object, either positive or negative.

Polymer: A chemical compound formed by polymerization.

Polyvinyl Chloride (PVC): A plastic material commonly used for pipe and plumbing.

Pool: A water receptacle used for swimming or bathing, designed to accommodate more than one person.

Portland Cement: One variety of cement and the basis of concrete and mortar.

Post-and-Beam Construction: A type of wood frame construction using timber for the structural support.

Post-Tensioning: The application of stretching steel cables embedded in a concrete slab to aid in strengthening the concrete.

Potable Water: Water which is satisfactory for drinking, cooking, etc.

Power: A basic unit of electrical energy, measured in watts.

Power Factor: The ratio of true power (kW) to apparent power (kVA).

Precipitation: The total measurable supply of water received directly from clouds as snow, rain, etc, expressed in inches.

Prehung: Refers to doors or windows that are already mounted in a frame and are ready for installation.

Pressure Treatment: Impregnating lumber with a preservative chemical under pressure in a tank.

Prestressed Concrete: Concrete in which the steel is tensioned (stretched) and anchored to compress the concrete.

Primary Winding: The input side of a transformer.

Primer: The first coat of paint or glue when more than one coat will be applied.

Private Sewage Disposal System: A septic tank with the effluent discharging into a subsurface disposal field and/or one or more seepage pits.

Private Sewer: A sewer not directly controlled by public authority.

Private Use: Plumbing fixtures in residences, apartments, private bathrooms in hotels and hospitals, rest rooms in commercial establishments containing restricted-use single fixture or groups of single fixtures, etc. where the fixtures are intended for use of family or individual.

Public Sewer: A common sewer directly controlled by public authority.

Public Use: Applies to locked and unlocked bathrooms used by employees, occupants, or patrons in any premises.

Purlin: A horizontal framing member spanning between rafters.

Putrefacation: Biological decomposition of organic matter with the production of foul-smelling products.

Q

Quarry Tile: An unglazed clay or shale flooring material produced by extrusion.

Quick Set: A fast-curing cement plaster.

R

Rabbet: A groove cut in or near the edge of a piece of lumber to fit the edge of another piece.

Raceway: Any partially or totally enclosed container for placing electrical wires.

Rafter: In sloped roof construction, the framing member extending from the ridge or hip to the top plate.

Reactance (X): The measure of inductance of a circuit, measured in ohms.

Rebar: A reinforcement steel rod in a concrete footing.

Receptacle: An electrical outlet to which an electrical device may be connected by means of a plug.

Receptor: A plumbing fixture that receives the discharge from indirect waste pipes that is made and located to be cleaned easily.

Reduced-Size Vent: Dry vents which are smaller than those allowed by codes.

Reducer: A pipe fitting with inside threads that are larger at one end than at the other.

Register: A grille used to cover an air duct opening.

Reglet: A long narrow slot in concrete to receive flashing or to serve as anchorage.

Relay: A device that controls one electrical circuit by opening and closing the contacts in another circuit.

Relief Valve: Designed to open automatically to relieve excess pressure.

Relief Vent: A vent designed to provide circulation of air between drainage and vent systems or to act as an auxiliary vent.

Residual Pressure: Pressure remaining in a system while water is being discharged from outlets.

Resilient Flooring: Flooring made of plastics rather than wood products.

Resistance (R): The opposition to the flow of current in an electrical circuit, measured in ohms.

Resistance-Weld Pipe: Pipe made by bending plate into circular form and passing current through to obtain a welding heat.

Resistive Circuit: A circuit containing resistive loads such as heating elements.

Resonance (f): When the inductive reactance (X_L) equals capacitive reactance (X_C) in a circuit, measured in hertz (Hz).

Return Offset: A double offset installed to return a pipe to its original alignment.

Revent Pipe: That part of a vent pipe line which connects directly with an individual waste or group of wastes, underneath or back of the fixture, and extends to the main or branch vent pipe.

R Factor: The numerical rating given any material that is able to resist heat transfer for a specific period of time.

Ridge: The highest point on a sloped roof.

Ridge Board: A horizontal member that connects the upper ends of the rafters on one side to the rafters on the opposite side.

Right-of-Way Line: A line on the side of a road marking the limit of the construction area and, usually, the beginning of private property.

Rim: An unobstructed open edge of a fixture.

Rise: In roofing, rise is the vertical distance between the top of the double plate and the center of the ridge board; in stairs, it is the vertical distance from the top of a stair tread to the top of the next tread.

Riser: A water supply pipe which extends vertically one full story or more to convey water to branches or fixtures; a vertical pipe used for fire protection to elevations above or below grade.

Riser: The vertical part at the edge of a stair.

Rolling Offset: Same as offset, but used where the two lines are not in the same vertical or horizontal plane.

Roll Roofing: A type of built-up roofing material made of rag, paper, and asphalt.

Roof Drain: A drain installed to remove water collecting on the surface of a roof and to discharge it into a leader.

Roof Jack: The sheet metal device placed around a pipe projecting through the roof to prevent moisture.

Roof Pitch: The ratio of total span to total rise expressed as a fraction.

Rotary Convertor: A type of phase convertor.

Rotary-Pressure Joint: A joint for connecting a pipe under pressure to a rotating machine.

Roughing-In: The installation of all parts of a plumbing system which can be completed prior to the installation of fixtures; this includes drainage, water supply, and vent piping.

Rough Opening: A large opening made in a wall frame or roof frame to allow the insertion of a door or window.

RS: Reference Stake, from which measurements and grades are established.

Run: A length of pipe made up of more than one piece; a portion of a fitting having its ends in line.

Run: The shortest horizontal distance measured from a plumb line through the center of the ridge to the outer edge of the plate.

R Value: The unit that measures the effectiveness of insulation; the higher the number, the better the insulation qualities.

S

Saddle: A small, gable-type roof constructed between a vertical surface such as the chimney and a sloped roof.

Saddle Flange: A flange curved to fit a boiler or tank and to be attached to a threaded pipe.

Sand Cone Test: A test for determining the compaction level of soil, by removing an unknown quantity of soil and replacing it with a known quantity of sand.

Sand Filter: A water treatment device for removing solid or colloidal material.

Sanitary Sewer: A conduit or pipe carrying sanitary sewage.

Saturated Steam: Steam at the same temperature as water boils under the same pressure.

Scabs: Boards used to join the ends of a girder.

Scale: A measuring device with graduations for laying off distances. Also: the ratio of size that a structure is drawn, such as $\frac{1}{4}$" × 1'-0" which is $\frac{1}{48}$ size.

Schedule: A list of details or sizes for building components, such as doors, windows, or beams.

Schematic: A one-line drawing for electrical circuitry or isometric plumbing diagrams.

Scissors Truss: A truss constructed to the roof slope at the top chord with the bottom chord designed with a lower slope for interior vaulted or cathedral ceilings.

Scraper: A digging, hauling, and grading machine having a cutting edge, a carrying bowl, a movable front wall, and a dumping mechanism.

Scratch Coat: First coat of plaster placed over lath in a three-coat plaster system.

Screed: A template to guide finishers in leveling off the top of fresh concrete. Screeding is "rough leveling."

Screwed Flange: A flange screwed on a pipe.

Screwed Joint: A pipe joint consisting of threaded male and female parts.

Scupper: An opening in a parapet wall attached to a downspout for water drainage from the roof.

Scuttle: Attic or roof access with cover or door.

Sealant: A material used to seal off openings against moisture and air penetration.

Seamless Pipe: Pipe or tube formed by piercing a billet of steel and then rolling.

Secondary Winding: The output side of a transformer.

Section: A vertical drawing showing architectural or structural interior design developed at the point of a cutting-plane line on a plan view; the section may be transverse—the gable end—or longitudinal—parallel to the ridge.

Seepage Pit: A lined excavation in the ground which receives the discharge of a septic tank and the effluent seeps through its bottom and sides.

GLOSSARY *(cont.)*

Seismic Design: Construction designed to withstand earthquakes.

Septic System: A waste system that includes a line from the structure to a tank and a leach field.

Septic Tank: Receives the discharge of a drainage system so as to separate solids from liquids and digest organic matter through a period of retention.

Series Circuit: A circuit that has only one current path.

Service Fitting: A street ell or tee with male threads at one end and female at the other.

Service Pipe: A pipe connecting water or gas mains with a building.

Set: Same as offset, but used where the connected pipes are not in the same vertical or horizontal plane.

Setback: In a pipe bend, the distance measured back from the intersection of the center lines to start of bend.

Setback: The distance from the property boundaries to the building location.

Sewage: Any liquid waste containing animal, vegetable, or chemical wastes in suspension or solution.

Sewage Ejector: A mechanical device or pump for lifting sewage.

Shakes: Shingles made of hand-split wood, in most cases western cedar.

Shear Wall: A wall construction designed to withstand shear pressure caused by wind or earthquake.

Sheathing: The outside layer of wood applied to studs to close up a house or wall; also used to cover the rafters.

Sheepsfoot Roller: A compacting roller with feet expanded at their outer tips used in compacting soil.

Shelf Angles: Structural angles which are bolted to a concrete wall to support brick work, stone, or terra cotta.

Shoring: Temporary support made of metal or wood, used to support other components.

Short Circuit: An undesired path for electrical current.

Short Nipple: A nipple whose length is longer than a close nipple.

Shoulder Nipple: Halfway between the length of a close nipple and a short nipple.

Shunt: Denotes a parallel connection.

Siamese: A hose fitting for combining the flow from two or more lines into a single stream.

Side Vent: A vent connected to a drain pipe through a fitting at an angle not greater than 45° to the vertical.

Sill: A piece of wood that is anchored to the foundation.

Sill Cock: A hose bibb.

Single-Phase Power: One of the three alternating currents in a circuit.

Sinker Nail: A nail for laying subflooring.

Size: Size is a special coating used for walls before wallpaper is applied.

Skewed: At an angle other than 99°.

Slab: A flat area of concrete such as a floor or drive.

Slab-on-Grade: The foundation construction for a structure with no crawl space or basement.

Sleeper: Wood strips laid over or embedded in a concrete floor for attaching a finished floor.

Sleeve Weld: Butting two pipes together and welding a sleeve over the outside.

Slip-On Flange: A flange slipped over the end of the pipe and then welded.

Sludge: The accumulated suspended solids of sewage deposited in tanks, beds, or basins and mixed with water to form a semiliquid.

Slump: The consistency of concrete at the time of placement.

Socket Weld: A joint made by use of a socket-weld fitting which has a prepared female end for insertion of the pipe to which it is welded.

Soffit: A covering for the underside of the overhang of a roof.

Soil Pipe: Any pipe which conveys the discharge of water closets, urinals, or fixtures to a building drain or sewer.

Solder Joint: A method of joining tube by use of solder.

Solenoid: An electromagnet with a movable iron core.

Soleplate: A 2×4 or 2×6 used to support studs in a horizontal position; it is placed against the flooring and nailed into position onto the subflooring.

Solid-Core Door: A flush door having an interior core of solid wood blocks glued together and an exterior of finished veneer paneling or other material, such as hardboard.

GLOSSARY *(cont.)*

Span: The horizontal distance between exterior bearing walls in a transverse section.

Spandrel Beam: The beam in an exterior wall of a structure.

Spandrel Wall: The portion of a wall above the head of a window and below the sill of the window above.

Special Wastes: Wastes which require some special method of handling, such as corrosion-resistant piping, sand, oil or grease interceptors, condensers, or other pretreatment facilities.

Specifications: The written instructions detailing the requirements for a project.

Specs: Short for Specifications. The written directions and detailed instructions which are used with the blueprints.

Spiral Pipe: Pipe made by coiling a plate into a helix and riveting or welding the edges.

Split-Phase Motor: A single-phase AC motor that has a running and a starting winding.

Split-Wired Receptacle: A receptacle that has the metal tap removed between the hot terminals.

Spoil Site: Area used to dispose of unsuitable or excess excavation material.

Spreader: Brace used across the top of concrete forms.

Sprinkler System: An integrated system of underground and overhead piping designed in accordance with fire protection standards.

Sprinkler System Classification:
1. Wet-pipe systems
2. Dry-pipe systems
3. Pre-action systems
4. Deluge systems
5. Combined dry-pipe and pre-action systems

Square: Refers to a roof-covering area; a square consists of 100 square feet.

Stack: The vertical main of a system of soil, waste, or vent piping extending through one or more stories.

Stack Group: The location of fixtures in relation to the stack.

Stack Vent: The extension of a soil or waste stack above the highest horizontal drain connected to the stack; also known as a waste or soil vent.

Stain: A paintlike material that imparts a color to wood.

Stainless Steel Pipe: An alloy steel pipe with corrosion-resisting properties.

Standard Pressure: Formerly used to designate cast-iron flanges, fittings, valves, etc., suitable for a maximum working steam pressure of 125 psi.

Standpipe: A vertical pipe generally used for the storage and distribution of water for fire extinguishing purposes.

Standpipe System: An arrangement of piping, valves, hose connections, and equipment installed in a structure with the hose connections located in such a manner that water can be discharged in streams or spray patterns for extinguishing fires.

Stepped Footing: A footing that may be located on a number of levels.

Stool: The flat shelf that rims the bottom of a window frame on the inside of a wall.

Stop Valve: The control of water supply to a single fixture.

Storm Sewer: A sewer used for conveying rain water, surface water condensate, cooling water, or similar liquid wastes exclusive of sewage.

Story: The space between two floors of a building or between a floor and the ceiling above.

Strain: Change of shape or size of body produced by stress.

Street Elbow: An elbow with male thread on one end and female thread on the other.

Stress: Reactions within a body resisting external forces acting on it.

Stress Skin Panels: Large prebuilt panels used as walls, floors, and roof decks built in a factory and hauled to the building site.

String Line: A nylon line usually strung tightly between supports to indicate both direction and elevation; used in checking grades or deviations in slopes or rises.

Strip Flooring: Wooden strips that are applied perpendicular to the joists.

Strongbacks: Braces used across ceiling joints that help align, space, and strengthen joists for drywall installation.

Structural Steel: Heavy steel larger than 12 gauge, identified by shape.

Stucco: A type of masonary finish used on the outside of a building applied over a wire mesh.

Studs: The vertical boards (usually 2×4 or 2×6) that make up the walls of a building.

GLOSSARY (cont.)

Subfloor: A platform that supports the rest of the structure underlayment.

Subgrade: The uppermost level of material placed in embankments or left at cuts in the normal grading of a road bed.

Sub-Main Sewer: A branch sewer into which the sewage from two or more lateral sewers is discharged.

Subsoil Drain: A drain which receives only subsurface or seepage water.

Summit: The highest point of any area or grade.

Sump: A tank or pit which receives sewage or liquid waste located below the grade of the gravity system.

Sump Pump: A mechanical device for removing liquid waste from a sump.

Super: A continuous slope in one direction on a road.

Superheated Steam: Steam at a higher temperature than that at which water would boil under the same pressure.

Superstructure: Frame of the building, usually above grade.

Supervisory Switch: A device attached to the handle of a valve, which, when the valve is closed, will annunciate a trouble signal.

Supports: Devices for supporting and securing pipe and fixtures.

Swale: A shallow dip made to allow the passage of water.

Sway Brace: A piece of 2×4 or similar material used to temporarily brace a wall from wind until it is secured.

Swedes: A method of setting grades at a center point by sighting across the tops of three laths; two laths are placed at a known correct elevation and the third is adjusted until it is at the correct elevation.

Swing Joint: An arrangement of screwed fittings and pipe that provides for expansion.

Switch (Electrical): A device to start or stop the flow of electricity.

Swivel Joint: A joint employing a special fitting that is pressure-tight under movement.

Symbol: A pictorial representation of a material or component on a plan.

T

Tail Joist: A short beam or joist supported in a wall on one end and by a header on the other.

GLOSSARY *(cont.)*

Tail/Rafter Tail: That portion of a roof rafter extending beyond the plate line.

Tamp: To pack tightly; usually refers to making sand tightly packed or making concrete mixed properly in a form to get rid of air pockets.

Tangent: A straight line from one point to another, which passes over the edge of a curve.

Taping and Bedding: Refers to drywall finishing; the application of specially prepared tape to drywall joints; bedding means embedding the tape in the joint to increase strength.

Taps: Connecting points on a transformer coil.

T-Beam: Beam which has a T-shaped cross section.

Tee: A fitting, either cast or wrought, that has one side outlet at right angles to the run.

Tempered Water: Water ranging in temperature from 85°F (29°C) up to 110°F (43°C).

Tensile Strength: The maximum stretching of a piece of metal (rebar, etc.) before breaking; calculated in kps.

Tensioning: Pulling or stretching of steel tendons to reinforce concrete.

Termite Shield: Sheet metal placed in or on a foundation wall to prevent intrusion.

Terrazzo: A mixture of concrete, crushed stone, calcium shells, and/or glass, polished to a tile-like finish.

Texture Paint: A very thick paint that will leave a texture or pattern.

Thermal Ceilings: Ceilings that are insulated with batts of insulation to prevent loss of heat or cooling.

Thermal Protection: Refers to an electrical device which has inherent protection from overheating.

Thermostat: An automatic device controlling the operation of HVAC equip.

Three-Phase Power: A combination of three alternating currents in a circuit with their voltages displaced 120° or one-third of a cycle.

Tie: A soft metal wire that is twisted around a rebar or rod and chair to hold in place until concrete is poured.

Tied Out: The process of determining the fixed location of existing objects (manholes, meter boxes, etc.) in a street so that they may be uncovered and raised after paving.

GLOSSARY *(cont.)*

Toe of Slope: The bottom of an incline.

Top Chord: The topmost member of a truss.

Top Plate: The horizontal framing member fastened to the top of the wall studs; usually doubled.

Trailer Park Sewer: The horizontal piping of a drainage system which begins 2 feet downstream from the last trailer site connection, receives the discharge of the trailer site, and conveys it to a sewage disposal system.

Transformer: A device which uses magnetic force to transfer electrical energy from one coil of wire to another.

Transverse: Across the short dimension of an object or structure.

Trap: A fitting designed to provide a liquid seal which will prevent the back passage of air without significantly affecting the flow of waste water through it.

Trap Primer: A device or system of piping to maintain a water seal in a trap.

Trap Seal: The maximum vertical depth of liquid that a trap will retain.

Travel: See Offset.

Tread: The part of a stair on which people step.

Trimmer: A piece of lumber, usually a 2 × 4, that is shorter than the stud or rafter but is used to fill in where the longer piece would normally have been spaced except for the opening in the roof, floor or wall.

True Power (P_T): The actual power used in an electrical circuit, measured in watts.

Truss: A prefabricated sloped roof system incorporating a top chord, bottom chord, and bracing.

Turbulence: Any deviation from parallel flow in a pipe due to rough inner wall surfaces, obstructions, etc.

Typical (Typ): This term, when associated with any dimension or feature, means the dimension or feature applies to the locations that appear to be identical in size and shape unless otherwise noted.

U

Underground Piping: Piping in contact with the earth below grade.

Underlayment: Also known as subfloor; used to support the rest of the building; also refers to the sheathing used to cover rafters.

Unfaced Insulation: Insulation which does not have a facing or plastic membrane over one side of it.

Union Ell: An ell with a male or female union at one end.

Union Joint: A pipe coupling, usually threaded, which permits disconnection without disturbing other sections.

Union Tee: A tee with a male or female union at one end of the run.

Upstream: Referring to a location in the direction of flow before reaching a reference point.

V

Vacuum: Any pressure less than that exerted by the atmosphere.

Vacuum Breaker: A backflow preventer.

Vacuum Relief Valve: A device to prevent excessive vacuum in a pressure vessel.

Valley: The area of a roof where two sections come together and form a depression.

Valley Rafters: A rafter which extends diagonally from the plate to the ridge at the line of intersection of two roof surfaces.

Vapor Barrier: A moisture barrier.

Veneer: A thin layer or sheet of wood.

Veneered Wall: A single-thickness (one-wythe) masonry unit wall with a backup wall of frame or other masonry; tied but not bonded to the backup wall.

Vent: Usually a hole in the eaves or soffit to allow the circulation of air over an insulated ceiling; usually covered with a screen.

Vent, Loop: Any vent connecting a horizontal branch or fixture drain with the stack vent of the originating waste or soil stack.

Ventilation: The movement of air through a building; may be done naturally through doors and windows or mechanically by fans.

Vent Stack: A system of pipes used for air circulation to prevent water from being suctioned into the traps in a waste disposal system.

Vertical Pipe: Any pipe or fitting installed in a vertical position or which makes an angle of not more than 45° with the vertical.

Vitrified Clay Tile: A ceramic tile fired at a high temperature

GLOSSARY *(cont.)*

to make it very hard and waterproof.

Vitrified Sewer Pipe: Conduit made of fired and glazed earthenware.

Void: Vacant space between material, such as a space in a column.

Volt (E) or (V): The unit of measurement of electrical pressure (force).

Voltage Drop: Voltage reduction due to resistance.

W

Waler: A 2x piece of lumber installed horizontally to formwork to give added stability and strength.

Waste Pipe: Discharge pipe from any fixture, appliance, or appurtenance in connection with a plumbing system which does not contain fecal matter.

Water-Cement Ratio: The ratio of the weight of water to cement.

Water Conditioner: Treats a water supply to change its chemical content or remove suspended solids by filtration.

Water-Distributing Pipe: A pipe which conveys potable water from a building supply pipe to the plumbing fixtures.

Water Hammer: The noise and vibration which develop in a piping system when a column of noncompressible liquid flowing through a pipe line at a given pressure and velocity is abruptly stopped.

Water Hammer Arrester: A device designed to provide protection against excessive (hammering).

Water Main: The water supply pipe for public or community use.

Waterproofing: Preferably called moisture protection; materials used to protect below- and on-grade construction from moisture penetration.

Water Riser: A water supply pipe which extends vertically one full story or more.

Water-Service Pipe: The pipe from a water main or other source of water supply to the building served.

Water Supply System: The building supply pipe, the water-distributing pipes, and the necessary connecting pipes, fittings, control valves, and all appurtenances carrying or supplying potable water in or adjacent to the building or premises.

Water Table: The amount that is present in any area.

Watt (W): The measure of electrical power.

Weep Holes: Small holes in a wall to permit water to exit from behind.

Welded-Wire Fabric (WWF): A reinforcement used for horizontal concrete.

Welding-End Valves: Valves with ends tapered and beveled for butt welding.

Welding Fittings: Beveled for welding to pipe.

Wet Vent: A vent which also serves as a drain.

Winder: Fan-shaped steps that allow a stairway to change direction without a landing.

Wind Lift (Wind Load): The force exerted by the wind against a structure.

Window Apron: The flat part of the interior trim of a window located directly beneath the window stool.

Window Stool: The flat, narrow shelf which forms the top member of the interior trim at the bottom of a window.

Windrow: The spill-off from the ends of a dozer or grader blade which forms a ridge of loose material.

Wiped Joint: A lead pipe joint in which molten solder is poured after scraping and fitting the parts together.

Working Drawings: A set of drawings which provide the necessary details and dimensions to construct the object. May include specs.

Wrought Iron: Iron refined to a plastic state in a puddling furnace.

Wrought Pipe: Refers to both wrought steel and wrought iron.

Wye (Y): A fitting that has one side outlet at any angle other than 90°

Wye Connection: Has one end of each coil connected together and the other end open for connections.

Wythe: A continuous masonry wall width.

X

X Brace: Cross brace for joist construction.

Y

Yoke Vent: A pipe connecting upward from a soil or waste stack to a vent stack for the purpose of preventing pressure changes in the stacks.

CHAPTER 10
Abbreviations

ABBREVIATIONS

&	And	AFF	Above finished floor
∠	Angle	AFG	Above finished grade
@	At	AFM	Terminal air flow module
°C	Degrees Celsius	AFS	Air flow station
°F	Degrees Fahr.	Ag	Silver
ø	Diameter	AGGR	Aggregate
"	Ditto, inch	AH, AHU	Air-Handling unit
'	Foot, feet	AL, ALUM	Aluminum
#	Number, pound	ALM	Alarm
%	Percent	ALT	Alternate
A	Air, amperes; ammeter, area, air line	AM	Ammeter
		AMB	Ambient
AB	Anchor bolt	AMP	Ampere
ABC	Above ceiling	AMT	Amount
ABS	Absolute	AP	Access panel
AC	Air chamber, alternating current	APOTH	Apothecary
		APPROX	Approximate
A/C	Air conditioning	ARM	Armature
ACC	Air-cooled cond.	ARR	Arrangement
ACCU	Air-cooled condensing unit	ASPH	Asphalt
		ATC	At ceiling or automatic temp. control
ACOUS	Acoustical		
ACP	Asbestos cement pipe	ATM	Atmosphere
		Au	Gold
ACT	Acoustical ceiling tile	AUTO	Automatic
ACU	Packaged air conditioning unit	AUX	Auxiliary

AV	Acid-resistant vent, air vent	B & S	Bell-and-spigot pipe
AVG	Average	BSMT	Basement
AW	Acid-resistant waste	BT	Bathtub
AWG	American wire gauge	BTU	British thermal unit
B	Bidet, Bathroom	BUS	Busway
BBD	Boiler blow-down	BUZ	Buzzer
BBL	Barrel(s)	BV	Butterfly valve
BBR	Baseboard radiation	C	Condensate line, Celsius, centigrade
BD	Board	C or COND	Conduit
BD FT (BF)	Board foot	C to C	Center to center
BEV	Beveled	CA	Compressed air
BF	Boiler feed, bottom of footing	CAB	Cabinet
		CAL	Calorie
BFP	Boiler feed pump	CAP	Capacitor, Capacity
BFS	Boiler feed set	CB	Catch basin, circuit breaker
BHP	Boiler, brake horsepower	CC	Center to center, cubic centimeter, cooling coil
BLDG	Building		
BLK	Black, block	CCW	Counterclock-wise
BLKG	Blocking	CD	Ceiling diffuser, condensate drain
BLR	Boiler		
BM	Beam, bench mark, board measure	CEM	Cement
		CER	Ceramic
BOD	Bottom of duct	CF	Charcoal filter, chemical feed, cubic foot (feet)
BOP	Bottom of pipe		
BOT	Bottom		
BP	Back pressure, base plate, blueprint	CFH	Cubic feet per hour
		CFM	Cubic feet per minute
BR	Bedroom, brass		
BRG	Bearing	CFP	Chemical feed pump
BRK	Brick		
BRKT	Bracket	CFU	Chemical feed unit
BRZ	Bronze	CH	Chiller
BS	Bar sink	CHAM	Chamfer
		CHR	Chilled water return

ABBREVIATIONS *(cont.)*

CHS	Chilled water supply	CRW	Chemical-resistant waste
CI	Cast iron	CSG	Clean stem generator
CIP	Cast or concrete-in-place	CSK	Countersink
CIR	Circuit	CT	Ceramic tile, cooling tower, current transformer
CIR BKR	Circuit breaker		
CIRC	Circular	CTR	Center
CISP	Cast-iron soil pipe	CU	Cubic, Copper
CJ	Ceiling joist, control joint	CU FT (ft^3)	Cubic foot (feet)
CKT	Circuit (elec)	CU FT.	Cubic feet
CL	Center line, closet	CU IN (in^3)	Cubic inch(es)
CLG	Ceiling	CU IN.	Cubic inches
CLK	Caulk	CU YD (yd^3)	Cubic yard(s)
CLR	Clear	CV or CKV	Check valve
CM	Centimeter	CW	Cold water, clockwise
CM^2	Square centimeter	CWM	Clothes washing machine
CM^3	Cubic centimeter		
CMU	Concrete masonry unit	CWR	Cold water riser
		CWS	Cold water supply
CO	Cleanout (plum)	d	Pennyweight
COL	Column	D	Diameter, drain line, drain; deep
COM	Common		
COMP	Composition	DA	Deaerator
CONC	Concrete; concentric	DAC	Door air curtain
CONN	Connect (ion)	DB	Decibel, dry bulb
CONST	Construction	DC	Direct current (elec.), duct coil
CONT	Continuous		
CONTR	Contractor	DDC	Direct digital control
CP	Candle power, concrete pipe, condensate pump	DEG	Degree
		DEH	Dehumidifier
		DET	Detail
CPS	Cycles per second	DF	Drinking fountain
CR	Ceiling register, control relay	DH	Double hung
		DIA, ø	Diameter
CRAC	Computer room air conditioning	DIAG	Diagonal
		DIM	Dimension, dimmer

DISC	Disconnect	**EMF**	Electromotive force
DIV	Division	**EMT**	Electric metallic tubing
DM	Decimeter		
DM²	Square decimeter	**ENAM**	Enamel
DM³	Cubic decimeter	**ENCL**	Enclosure
DMPR	Damper	**ENGR**	Engineer
DN	Down	**ENT**	Entrance
DO	Ditto (same as)	**EQUIP**	Equipment
DP	Double-pole, duplicate	**ESP**	External static pressure
DPDT	Double-pole double-throw	**EST**	Estimate
		ET	Expansion tank
DPG	Damp proofing	**EUH**	Electric unit heater
DR	Drain, dining Rm	**EVAP**	Evaporator
DS	Downspout	**EW**	Each way
DUC	Dust collector	**EWC**	Electric water cooler
DW	Dishwasher, dry wall	**EWT**	Entering water temperature
DWG	Drawing		
DWL	Dowel	**EXCAV**	Excavate
DWV	Drain, waste, and vent system	**EXCL**	Exclude
		EXH	Exhaust
E to C	End to center	**EXP**	Expansion
EA	Each	**EXST**	Existing
EAT	Entering air temperature	**EXT**	Exterior, external
		EXTN	Extension
EDR	Equivalent direct radiation	**F**	Fahrenheit, fire line, fan, forward; fast, field; filter
EER	Energy efficiency ratio		
		FAB	Fabricate
EF	Exhaust fan	**FAM RM**	Family room
EFF	Efficiency	**FAO**	Finish all over
EJ	Expansion joint	**FB**	Foot bath
EL	Elevation	**F BRK**	Fire brick
ELB	Elbow	**FC**	Flexible connection
ELEC	Electrical	**FCO**	Floor cleanout
ELEV	Elevation; elevator	**FCU**	Fan coil unit
EMER	Emergency	**FD**	Floor drain

FDC	Fire dept. conn.	**FREQ**	Frequency
FDN	Foundation	**FRG**	Furring
FDR	Feeder	**FS**	Float or flow switch, floor sink
FDW	Feed water		
FEC	Fire extinguisher cabinet	**FSP**	Fire standpipe
		F&T	Float & thermostatic
FF	Finish floor	**FT**	Foot, flash tank
FG	Finish grade	**FTG**	Fitting, Footing
FHC	Fire hose cab.	**FTS**	Foot switch
F & I	Furnish/install	**FU**	Fixture unit
FIG	Figure	**FURN**	Furnish, furnace
FIN	Finish	**FV**	Flush valve
FIN FLR	Finish floor	**FWD**	Forward
FIN GRD	Finish grade	**FWH**	Frostproof wall hydrant
FIX	Fixture		
FL	Flow meter, flashing	**FX GL (FX)**	Fixed glass
FL, FLR	Floor	**FX WDW**	Fixed window
FLA	Full-load amps	**G**	Gas, gram, gate, gas line, green
FLC	Full-load current		
FLEX	Flexible	**GA**	Gauge
FLG	Flange, flooring	**GAL**	Gallon(s)
FLT	Full-load torque	**GALV**	Galvanize(d)
FLUOR	Fluorescent	**GAR**	Garage
FM	Fire line, flow meter	**GB**	Grade beam
FMB	Filter mixing box	**GC**	General contractor, glycol cooler
FO	Fuel oil		
FOB	Free-on-board	**GD**	Ground
FOM	Face of masonry	**GEN**	Generator
FOP	Fuel oil pump	**GI**	Galvanized iron
FOS	Face of stud; flush on slab	**GL**	Glass, Glazed
		GL BLK	Glass block
FOV	Flush-out valve	**GLB, GLU-LAM**	Glue-laminated beam
FP	Fire plug, fireplace		
FPM	Feet per min.	**GLV**	Globe valve
FPS	Feet per sec.	**GND**	Ground
FR	Frame	**GPD**	Gallons per day
		GPH	Gallons per hour

ABBREVIATIONS *(cont.)*

GPM	Gallons per min.	HP	Heat pump, horsepower
GPS	Gallons per sec.	HR	Hour
GR	Grain	HRC	Heat-recovery coil
GRAN	Granular	HRU	Heat-recovery unit
GRD	Grade, ground	HTD	Heated
GRTG	Grating	HTR	Heater
GS	Glass sink	HV	Heating and ventilating unit
GV	Gate valve		
GWB	Gypsum wall board	HVAC	Heating, ventilating, and air conditioning
GWH	Gas water heater		
GY	Gray	HW	Hot water
GYP	Gypsum	HWH	Hot water heater
H	Hall, humidifier, transformer on primary side	HWR	Hot water return or hot water riser
		HWT	Hot water tank
H_2O	Water	Hz	Hertz
HB	Hose bibb	I	Current
HC	Heating coil	IB	Inlet basin
Hd/HD	Head	IC	Integrated circuit
HDR	Header	ID	Inside diameter
HDW	Hardware	IF	Inside face
HDWD	Hardwood	IN	Inch(es)
HE	Heat exchanger	INCAND	Incandescent
HEF	High-efficiency filter	INCL	Include
HEPA	High-efficiency particulate air filter	INCR	Increase
		INHg	Inches of mercury
HEX.	Hexagonal	INSUL	Insulation
Hg	Mercury	INT	Interior, Intermediate
HG	Hose gate, hot gas	INTL	Internal
HGT/HT	Height	INV	Invert
HM	Hollow metal	IPS	Iron pipe size
HMD	Humidity	ITB	Inverse time breaker
HORIZ	Horizontal	ITCB	Instantaneous trip circuit breaker
		IV	Indirect vent
		IW	Indirect waste

ABBREVIATIONS *(cont.)*

J	Joule, Junction	LEV/LVL	Level
J, JST	Joist	LFM	Laminar-flow module
JB	Junction box		
JT	Joint	LG	Long
K	Kelvin, kilo; cathode, kip (1,000), kitchen	LGTH	Length
		LH	Left hand
KAL	Kalamein	LIN	Lineal
KG	Kilogram	LIN FT (LF)	Lineal foot (feet)
KL	Kiloliter	LIQ	Liquid
KM	Kilometer	LITE/LT	Light (pane)
KM²	Square kilometer	LOG	Logarithm
KP	Kick plate	LP	Low pressure
KPA	Kilopascal	LPG	Liquefied petroleum gas
KS	Kitchen sink		
KVA	Kilovolt-amperes	L & PP	Light and power panel
KW	Kilowatt		
KWH	Kilowatt-hour	LR	Living room
L	Length or liter, left, line, load	LRA	Locked rotor amps
		LRC	Locked rotor current
L or LAV	Lavatory	LT	Laundry tray
LAB	Laboratory	LTH	Lath
LAD.	Ladder	LVL	Level
LAM.	Laminated	LVR	Louver
LAT	Lateral, Leaving air temperature	LWT	Leaving water temperature
LAU	Laundry	M	Meter, motor; motor starter contacts, (1000)
LB	Light beam		
LB-FT	Pounds per feet		
LB-IN	Pounds per inch	M²	Square meter
LB.	Pound	MAGN	Magnesium
LBF	Pound-force	MAH	Makeup air heater
LBR	Labor, lumber	MAN	Manual
LBS	Pounds	MAS	Masonry
L CL	Linen closet	MAT'L	Material
LDG	Landing, leading	MAV	Manual air vent
LDR	Leader	MAX	Maximum

MB	Mixing box	**MS**	Manual starter
MBF/MBM	Thousand board feet, thousand board measure	**MTD**	Mounted
		MU	Makeup
		MULL	Mullion
MBH	Thousand British thermal units per hour	**N**	Nirth, north
		NA	Not applicable
MC	Medicine cabinet	**NC**	Normally closed
MCM	Thousand circular mils	**NEG**	Negative
		NF	Near face
MDP	Main dist. panel	**NIC**	Not in contract
MECH	Mechanical	**N-M**	Newton-meter
MED	Medium	**NO**	Normally open
MET.	Metal	**NO.**	Number
MEZZ	Mezzanine	**NOM**	Nominal
MFR	Manufacturer	**NOR**	Normal
MG	Milligram	**NPS**	Nominal pipe size
MGD	Million gallons per day	**NTDF**	Nontime-delay fuse
		NTS	Not to scale
MH	Manhole	**O**	Oxygen, orange
MI	Malleable iron	**OA**	Outside air
MIN	Minimum or minute	**O/A**	Overall (meas)
MISC	Miscellaneous	**OAI**	Outside air intake
MK	Mark (identifier)	**OBSC GL**	Obscure glass
ML	Milliliter	**OC (O/C)**	On center
MLDG	Molding	**OCPD**	Overcurrent protection dev.
MM	Millimeter	**OD**	Outside diameter
MM³	Cubic millimeter	**OED**	Open-end drain, Open-end duct
MN	Main		
MO	Momentary, masonry opening, motor operated	**OF**	Overflow
		OF	Outside face
		OFCI	Owner furnished contractor installed
MOD	Model, modular		
MOR	Mortar	**OFF**	Office
MOUNT	Mounting	**OH**	Overhead
MPT	Male pipe thread	**O/H**	Overhang

OL	Overloads	PLWD	Plywood
OPG	Opening	PNEU	Pneumatic
OPP	Opposite	PNL	Panel
OUT	Outlet	PORC	Porcelain
OV	Outlet velocity	PP	Pool piping
OVHD	Overhead	PR	Pair
OZ	Ounce	PREFAB	Prefabricate(d)
OZ/IN	Ounces per inch	PRESS	Pressure
P	Power, pump	PRI	Primary
PA	Pascal	PROP	Propeller
PAN	Pantry	PROP	Property, proposed
PAR	Parallel	PRV	Pressure-reducing valve
PASS	Passage	PSG	Pure steam generator
PAV	Paving		
PB	Pushbutton	PSI(G)	Pounds per square inch (gauge)
PC	Piece, pull chain		
PCR	Pumped condensate return	PTAC	Packaged terminal air conditioner
PD	Planter drain, pressure drop	PTN	Partition
		PUT	Pull-up torque
PERIM	Perimeter	PV	Plug valve
PERP	Perpendicular	PWR	Power
PF	Prefilter	QT	Quarry tile, quart
PG	Pressure gauge	QTY	Quantity
PH	Phase	R	Right, radius, Resistance; red, Risers
PHC	Preheat coil		
PL	Pilot light, plate, property line	RA	Return air
		RAD	Radius, radiator
PLAST	Plaster	RCP	Reinforced concrete pipe
PLAT	Platform		
PLBG	Plumbing	RD	Road; round; roof drain
PL GL	Plate glass		
PL HT	Plate height	REBAR	Reinforced steel bar
PLT	Plate (framing)	REC	Recessed
		RECEPT	Receptacle

RECOV	Recovery	**SB**	Sitz bath
RED	Reducer	**SC**	Self-closing, sill cock, steam coil
REF	Reference, refrigerator	**SCH/SCHED**	Schedule
REG	Register	**SCR**	Screen, screw
REINF	Reinforce(ment)	**SCUP**	Scupper
REQD	Required	**SD**	Smoke damper, storm drainage
REQ'D	Required		
RET	Retain(ing); return	**SDG**	Siding
REV	Reverse, revision	**SDL**	Saddle
RF	Roof	**SEC**	Second, secondary
RFG	Roofing (mat.)	**SECT**	Section
RGH	Rough	**SEL**	Select
RGH OPNG	Rough opening	**SENS**	Sensible
RH	Relief hood, right hand	**SEP**	Separate
		SEQ	Sequence
RIO	Rough-in opening	**SER**	Series
RL	Roof leader, refrigerant liquid	**SERV**	Service
		SEW	Sewer
RM	Room	**SF**	Service factor
RPM	Revolutions per minute	**SG**	Supply air grille
		SH	Shower, Sheet
RS	Refrigerant suction	**SH & RD**	Shelf and rod
RTN	Return	**SH HD**	Shower head
RV	Relief valve	**SHLP**	Shiplap
RWD	Redwood	**SHT**	Sheet
RWL	Rain water leader	**SHTHG**	Sheathing
S	Soil line, stretcher, switch; series; slow; south	**SHWR**	Shower
		SIM	Similar
		SJ	Steel joist
SA	Shock absorber, supply air	**SK**	Sink
		SL	Slate, Sleeve
SAD	Supply air diffuser	**SM**	Sheet metal
SAN	Sanitary	**SOL**	Solenoid
		SOV	Shutoff valve

ABBREVIATIONS (cont.)

SP	Swimming pool, single-pole, soil pipe (plumbing), static pressure, sump pump	SUCT	Suction
		SUR	Surface
		SUSP CLG	Suspended ceiling
		SV	Service, Safety valve
		SW	Service weight, Switch
SPAN	Spandrel		
SPD	Sump pump discharge	SWS	Service water
SPDT	Single-pole double-throw	S & W	Soil and waste
		SYM	Symbol, symmetric
SPEC	Specification	SYS	System
SPKR	Speaker	T	Tempered water, potable; or temperature, thermostat
SPLY	Supply		
SPR	Sprinkler		
SPST	Single-pole single-throw	TAN	Tangent
		TB	Top of beam
SQ	Square	T & B	Top and bottom
SQ FT	Square feet	TC	Terra cotta, top of concrete, top of curb
SQ FT (ft²)	Square foot		
SQ IN (in²)	Square inch(es)	TD	Temp. diff., Time delay
SQ YD (yd²)	Square yard(s)		
SR	Supply air register	TDF	Time-delay fuse
SS	Service sink, sanitary sewer, stainless steel	TEL	Telephone
		TEMP	Temperature
		TEMPL	Template
ST	Storm	TERR	Terrazzo
STA	Station	TF	Top of footing
STD	Standard	T&G	Tongue/groove
STIFF.	Stiffener	TH	Thermometer
STIR	Stirrup (rebar)	THK	Thick
STL	Steel	THR	Threaded
STM	Steam	TJ	Top of joist
STN	Stone	TM	Top of masonry
STOR	Storage	TMB	Telephone mounting board
STR	Straight, strainer		
STR/ST	Street	TOB	Top of beam
STRUCT	Structural		

TOC	Top of curb	**VCL**	Vacuum cleaning line
TOF	Top of footing	**VEL**	Velocity
TOIL	Toilet	**VERT**	Vertical
TOL	Top of ledger	**VEST**	Vestibule
TOP	Top of parapet	**VF**	Variable air volume box
TOS	Top of steel		
TP	Top of pier	**VIB**	Vibration
TR	Tread, transition	**VOL**	Volume
TRANS	Transformer	**VP**	Vitreous pipe
TRK	Track, truck	**VR**	Vent riser
TS	Tensile strength, time switch, top of slab	**VS**	Vent stack
		VSD	Variable-speed drive
		VTR	Vent through roof
TSP	Total static press.	**W**	Waste or waste line, watt; width, west, white
TV	Turning vanes		
TW	Top of wall		
TYP	Typical	**W/**	With
U or UR	Urinal	**WB**	Wet bulb
UF	Under floor, underground feeder (elec.)	**WC**	Water closet
		WCL	Water cooler
		WCO	Wall cleanout
UH	Unit heater	**WD**	Wood
UNFIN	Unfinished	**WDW**	Window
USE	Underground service entrance cable (electrical)	**WF**	Water or wash fountain, wall fin radiation
V	Vent line, valve, variable air, volume box, vent; volt	**W GL**	Wire glass
		WH	Water heater, weep hole
VA	Voltamps	**WHR**	Watt-hour
VAC	Vacuum	**WI**	Wrought iron
VAN	Vanity	**WM**	Washing mach. water meter
VAR	Varies		
VAV	Variable air volume	**W/O**	Without
VB	Vacuum breaker	**WP**	Weatherproof, work point
VC	Vitrified clay		
VCI	Vacuum cleaning inlet	**WR**	Washroom

ABBREVIATIONS *(cont.)*

WR BD	Weather-resistant board	**X**	Transformer on secondary side
WS	Weather strip	**XH**	Extra heavy
WT/WGT	Weight	**Y**	Wye, yellow
WV	Water valve	**YD**	Yard
WWF	Welded-wire fabric		

About The Author

Paul Rosenberg has an extensive background in the construction, data, electrical, HVAC and plumbing trades. He is a leading voice in the electrical industry with years of experience from an apprentice to a project manager. Paul has written for all of the leading electrical and low voltage industry magazines and has authored more than 30 books.

In addition, he wrote the first standard for the installation of optical cables (ANSI-NEIS-301) and was awarded a patent for a power transmission module. Paul currently serves as contributing editor for *Power Outlet Magazine*, teaches for Iowa State University and works as a consultant and expert witness in legal cases. He speaks occasionally at industry events.